BASIC PLUMBING TECHNIQUES

Created and designed by the editorial staff of ORTHO Books

Project Editor
Ken Burke

Writer
Ron Hildebrand

Designers
Craig Bergquist
Christine Dunham

Illustrator
Ron Hildebrand

Photographer
Fred Lyon

Photographic Stylist
Sara Slavin

Ortho Books

Publisher
Robert L. Iacopi

Editorial Director
Min S. Yee

Managing Editors
Anne Coolman
Michael D. Smith
Sally W. Smith

Production Manager
Ernie S. Tasaki

Editors
Jim Beley
Susan Lammers
Deni Stein

Design Coordinator
Darcie S. Furlan

System Managers
Christopher Banks
Mark Zielinski

Photographic Director
Alan Copeland

Photographers
Laurie A. Black
Richard A. Christman

Production Editors
Linda Bouchard
Alice Mace
Kate O'Keeffe

Asst. System Manager
William F. Yusavage

Chief Copy Editor
Rebecca Pepper

Photo Editors
Anne Dickson-Pederson
Pam Peirce

National Sales Manager
Garry P. Wellman

Sales Associate
Susan B. Boyle

Operations Director
William T. Pletcher

Operations Assistant
Gail L. Davis

Administrative Assistant
Georgiann Wright

Address all inquiries to
Ortho Books
Chevron Chemical Company
Consumer Products Division
Box 5047
San Ramon, CA 94583

First Printing in March 1982.

12 13 14
90 91 92 93

ISBN 0-89721-001-8

Library of Congress Catalog Card
Number 81-86182

Chevron Chemical Company
6001 Bollinger Canyon Road, San Ramon, CA 94583

Consultants:
Robert Beckstrom
Owner/Builder Center
Berkeley, CA

Thomas P. Konen
Professional Engineer
Bridgewater, NJ

Robert Wyly
Kensington, MD

Front Cover Photographer:
Fred Lyon

Typography:
Vera Allen Composition
Castro Valley, CA

Color Separations:
Colortech
Redwood City, CA

Copyediting:
Carol Westberg
San Francisco, CA

Front Cover—Plumbing pipes are available in various materials and sizes.

Back Cover—This book will guide you through everything from choosing and insulating pipes to fixing a faucet to plumbing a new bathroom.

Title Page—Some of the several tools most do-it-yourself plumbers may need.

BASIC PLUMBING TECHNIQUES

PLUMBING SYSTEMS

From fixing a faucet to putting in a shower, you can learn the steps one at a time. Start with a look at plumbing codes and permits and the tools you will need. Find out about your water meter, vents, traps, sump pumps, and more.

If you've hired a plumber lately, you know that it is an expensive proposition. If this book can help you handle even the simple plumbing problems yourself, it may save you a lot of money over the next few years. Even if you're not concerned about the plumber's fees, you may have other good reasons for learning to handle some basic jobs on your own—like the toilet that overflows on Sunday morning or the pipes you want to keep from freezing. Or maybe you just like knowing how to do things for yourself.

It is possible for you to do virtually all the necessary plumbing in your house. Maybe not right now, but with this book and the will to do a job, even the most complicated plumbing can soon be within your range. Even though the plumbing system of a house is made up of thousands of pieces, it is all put together one piece at a time. Anyone can cement two plastic fittings together—ask a ten-year-old child who builds model airplanes. And with a little practice, sweating copper tubing to a fitting is not only easy, it's fun.

Many of the procedures in this book can be followed exactly in a step-by-step manner. But since every house is different and each manufacturer makes fixtures differently, you must study and plan each project carefully before starting the work.

What Plumbing Involves

Plumbing in a house is a system of pipes, valves, tanks, fixtures, and appliances for the distribution, use, and disposal of water, sewage, and gas. Plumbing also includes the art or trade of installing and maintaining such a system.

Most homes have three separate plumbing systems: a water supply system, a wastewater or sewage system, and a gas supply system. Each of these systems has unique properties and problems, but they have enough similarities that as you work with one of them, you will be gaining experience that will apply to the others.

The water supply system brings potable water into your house, regulates its pressure, and distributes it to all the places you may want it. This water comes from a public or privately operated water main under the street or from a well, spring, or other private source of your own. This system usually has a means of heating some of the water for you, and it may have the means to purify or change the water in other ways to make it more useful to you.

The wastewater or sewage system collects and takes used water out of your house and delivers it to a public sewer system or your own seepage or septic system. This system also takes the fumes, odors, and gases from the sewage and vents them safely through the roof.

The gas supply system brings natural gas or propane into your home from gas-company pipes beneath the street or your own tank and distributes it to appliances throughout the house.

This book will show you all the plumbing that is hidden under the floors and within the walls of your home, how the pipes and fixtures are put together, how everything works, and how to repair, maintain, and even extend the systems. It will also show you how to remove old fixtures or appliances and install new ones. For each repair or maintenance job, it will tell you what materials and tools you'll need and give you step-by-step guidance, including simple drawings and diagrams.

Before you start any plumbing project, study each step carefully and be sure you understand it. Gather all the materials and tools you'll need before you begin. Then proceed in a methodical, step-by-step manner from the beginning to the end. Soon projects that looked hard at first will look much easier, and before you know it, you'll be tackling projects you never imagined doing yourself.

◀
Every fixture in your house that holds water is connected to the water supply and disposal systems.

BASIC PLUMBING TOOL SET

If you are starting from scratch with no tools at all, don't despair. You probably need fewer tools for plumbing than for any other kind of do-it-yourself work. Carpentry, auto mechanics, and gardening all require considerably more tools than plumbing does. And plumbing tools are, on the average, less expensive.

The entire basic tool set we recommend for emergency and minor repairs can be bought for less than $100. Even if you select very high-quality wrenches, screwdrivers, and pliers, you can probably get them all for around $125. Even if you are planning on doing only occasional emergency repairs, we recommend that you buy good tools. Get the best you can afford; or even better, get tools that cost a little more than you think you can afford. It will really pay off in the long run, both in the cost of replacement tools and in the frustration that broken tools and skinned knuckles can bring.

The only tools and materials you really need to have on hand are those needed for emergency or minor repairs that crop up from time to time. The tools for additions or modifications can be selected and purchased or rented when you decide to do the job. Many plumbing tools—like screwdrivers, hammers, and pliers—are used for so many other things around the house that you should have them in your tool box anyway.

Emergencies do happen, and even if you don't think you are capable of doing anything yourself, we'll guarantee the tools will come in handy eventually. A guest or neighbor can often do wonders to prevent extensive damage if the proper tools are handy when an emergency occurs.

Here is the basic tool set that we recommend to every homeowner, whether or not he or she plans to do plumbing repairs.

Plunger. This tool is also called a plumber's helper, plumber's friend, or force cup. When it is worked up and down in the drain of a sink or toilet, the alternate pressure and suction can often dislodge clogs. There are two types. One has a funnel-shaped extension on the bottom, and the other is plain. It is desirable to have both types, but if you must make a choice, buy the one with a funnel-shaped extension. It works much better in toilets than the one without the extension. The extension can be retracted up into the cup if it is not needed or gets in the way.

Drain and trap auger. Also called a snake, this tool is used to dislodge clogs in traps and drains. The most common size is about 10 to 12 feet long and made of ¼-inch, tightly twisted steel wire with a corkscrewlike head called a gimlet. This type is very useful for clogs in sink traps and branch drain lines. A longer snake, 25 feet long with a ½-inch diameter cable, is useful for large clogs in main drains. You can usually rent a larger snake, with an electric motor or hand-crank and gears, at a plumbing supply or tool rental establishment.

Closet auger. This 4- to 5-foot auger with a curved, tubular handle is similar to a snake and works the same way. However, it is specially designed to unclog toilets without damaging the surface of the bowl. If you insert the curved end of the handle into the drain hole of the toilet before feeding the auger into the drain, the auger never rubs on the visible vitreous china surface.

Pipe wrenches. Older plumbers may call these Stillson wrenches. They are tooth-jawed, adjustable wrenches designed to grip pipe. You will need two of them: one 12 to 14 inches long and another 18 inches long. If you are working on ½-inch or ¾-inch pipe, hold the pipe with the larger one and do the turning with the smaller one. Do it the opposite way on 1- or 1½-inch pipe. That way you'll make the connections snug without the probability of tightening them too much and damaging the threads. If you use these wrenches on chrome-plated or brass pipes that are exposed, protect the finish with tape or rags.

Smooth-jaw adjustable wrenches. By this we mean monkey or crescent wrenches to use on nuts, bolts, and square or hexagonal fittings. For most plumbing purposes, you'll need wrenches at least 10 to 12 inches long. The jaws of smaller ones won't open far enough to hold pipe fittings.

Screwdrivers. You'll need two or three sizes of both slot and phillips screwdrivers.

Pliers. We recommend a regular pair of slip-jaw pliers and a larger pair of pump or channel-lock pliers.

Flashlight. A lot of plumbing is done in the deepest, darkest recesses of your house, and emergencies tend to happen at night. Always have a flashlight and extra batteries handy. You won't be sorry.

Trouble light. This is a light bulb, protected by a wire cage, on the end of a long extension cord. It has a hook so you can hang it near where you are working. It is much better than a flashlight when you are working for an extended period in a dark place.

Valve seat tool. Although we really can't call this a "basic" tool, if you live in an older house with lots of compression faucets, one of these will come in handy fairly often. It is used to smooth and repair worn or damaged valve seats on faucets. There are several types available, as shown in the drawings. We recommend the one with the reversible double cone guide as being the easiest to use, as well as the least likely to gouge the seat or leave burrs. The size of the seat varies with different faucets. You should probably have 3 or 4 different-sized cutters to be sure you have the proper one when the time comes.

Valve seat wrench. As long as you have a seat grinding tool, you might as well have a seat wrench, too. This is an inexpensive piece of bent metal with a hexagonal tool on one end and a square tool on the other. It is used to remove and replace valve seats that are beyond resurfacing.

Wire brush. A wire brush is useful to clean old pipe threads so you can use them again. It is good to have around for the clean-up part of other plumbing repairs and many other odd jobs around the house.

Measuring tape. Another tool that is indispensable when doing plumbing repairs and installations and very useful for many other household jobs is a self-retracting steel tape measure like carpenters use. Get one at least 12 feet long. A 25-foot one is even better.

Plumbing Tools

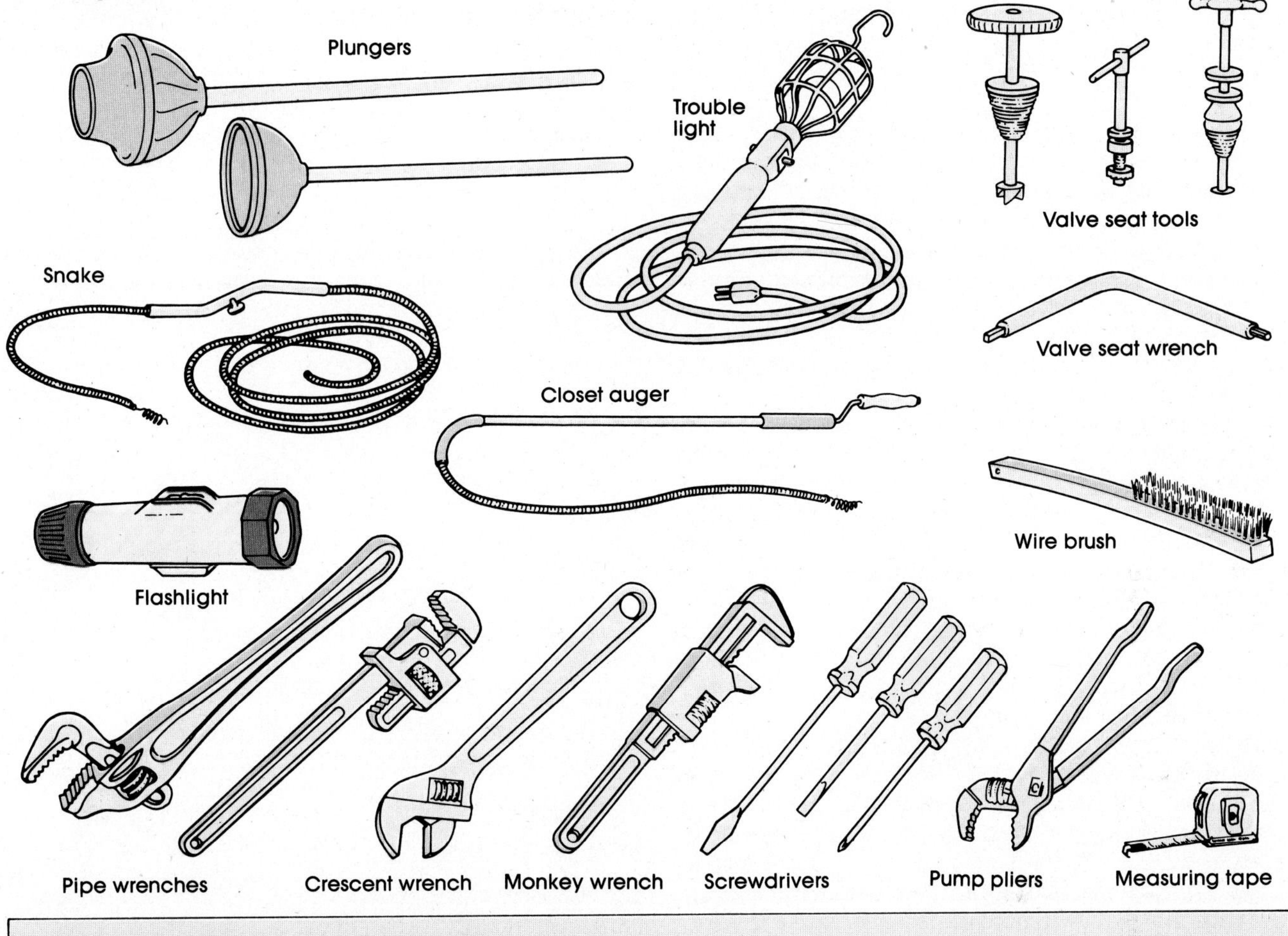

Plumbing Codes and Permits

The plumbing system in your house probably conforms to the building code in effect today. When you need to make emergency repairs, you do not need a permit. As long as your repairs restore the plumbing to its original condition or better, you have no problem.

All states, counties, and municipalities have adopted building codes to protect the health and safety of their residents. The codes also prevent unscrupulous builders and contractors from using inferior methods and materials in constructing, remodeling, and repairing your house. The details of codes vary from place to place. Materials, design, and methods of construction may vary due to local geography, experience, or tradition. Some localities are slow to adopt new materials or methods into their codes. Some places, for example, still do not allow plastic pipe to be used for water supply lines.

Building codes are usually administered and enforced by the building inspector of the state, county, or city in which you live. Even if no one can see what you are doing to the plumbing in your own basement, you must abide by all the rules and regulations of the code, just like the contractor who is building a new house. You are subject to all the same restrictions—and penalties—as the contractor if you don't get a permit when one is required or if your work is not up to the code.

It's not difficult to comply with the codes. It's just a matter of planning ahead and using the right materials and tools for the job. Whether or not your plumbing works the way it's supposed to and meets the code depends not so much on who does it and how many degrees or licenses he or she has, but on how the work is done.

This book should not be used as if it were a plumbing code. Even though the way we show things might vary only in minor ways from some codes, you should refer to your local code before making any changes or additions to your plumbing. The descriptions and drawings here should give you the basic knowledge necessary to understand and abide by the local code.

Permits are usually required whenever you make any changes or additions to your plumbing system. Repair or replacement of fixtures or appliances within the existing system, if it does not alter the system in any way, does not generally require a permit. If you are in doubt, ask at your local building inspector's office. Whether or not a permit is required, building inspectors and their staffs are usually very cooperative and helpful.

THE WATER YOU USE

Using Your Water Meter

Obviously, the primary use for a water meter is to let the water company know how much water you use so they can charge you for it. There are, however, a few other uses for your water meter.

Areas with a metered water supply can be managed more efficiently than those without water meters. The water company, as well as the city, county, state, and federal governments, can use water meter figures to determine how much water is being used by customers and how much is being wasted by evaporation or leaks in the system. They can determine how much is being used by various customer groups: farmers, industries, households, geographical areas, or any other group they need to know about.

This information is valuable in the intelligent planning of future water needs in any area. It allows planners to base their recommendations for new facilities on actual water usage rather than on guesses and estimates, which they would have to use if there were no meters. Many of the political controversies about water in New York City could probably be settled much more quickly and easily if that city had water meters to prove what was happening with the water after it entered the city.

Tracking water usage. You can use your meter for more than just "checking up" on the water company's billing. You can determine how much of your water is used for any one purpose, such as the garden or your teenager's 20-minute morning showers. Especially in times of water shortage or drought, knowing how much water you use for bathing, laundry, washing the car, or filling the swimming pool can aid you in planning and conservation.

To determine the amount of water used in any operation—taking a long shower, for instance—turn off all the water in the house and garden. Read your water meter, as described in the box, and write down the number. Then go take your shower. When you have finished, read the meter again. Subtract the first number from the second, and the answer will be the amount of water you used, usually in cubic feet. There are 7.5 gallons in a cubic foot, so to find the number of gallons you used, multiply the cubic feet by 7.5.

To find out how much your showers are costing you each month, look at your water bill to see how you are charged. Most bills list one or more fixed costs, like fire protection or capital improvements, and one cost based

Reading a Water Meter

In most areas, water is measured in cubic feet and you are charged for water on the basis of how many hundreds of cubic feet you use. Three types of meters are commonly used in the United States for home water consumption measurement. One of them has a direct digital readout that resembles the odometer on your automobile. This kind usually has a sweep needle that measures one cubic foot with each revolution. The one-cubic-foot dial is divided into tenths.

The second type has a large dial and also a sweep needle that indicates the passage of one cubic foot of water with each revolution. In addition, it has five small dials that all move clockwise. Each is divided into tenths, and each is labeled for the number of cubic feet it records with each revolution—10, 100, 1,000, 10,000, and 100,000. To read this kind of meter, look at the 100,000 dial first and note the smallest of the two numbers nearest the needle. Then note the reading of each of the other dials in descending order, the last digit coming from the 10 dial.

The most common type of water meter is the one with six dials. It operates and is read the same way as the one with five small dials except that the dials alternate clockwise and counterclockwise in their motion. Look carefully and be sure to note the *smaller* of the two numbers nearest the needle on each dial. The one-foot dial is usually not marked with ten divisions on this type of meter. This dial is mainly intended to show the passage of water, like when you are trying to detect a leak.

All of the meters shown here have the same reading, 371,940 cubic feet.

Most water companies charge by the hundreds of cubic feet of water, so they only read and record the first three digits from the meter. On your bill you will find something like this:

Prior reading	350
Present reading	371
Consumption in 100 cu. ft. (CCF)	21

You are then charged for 2,100 cubic feet of water at the prevailing rate.

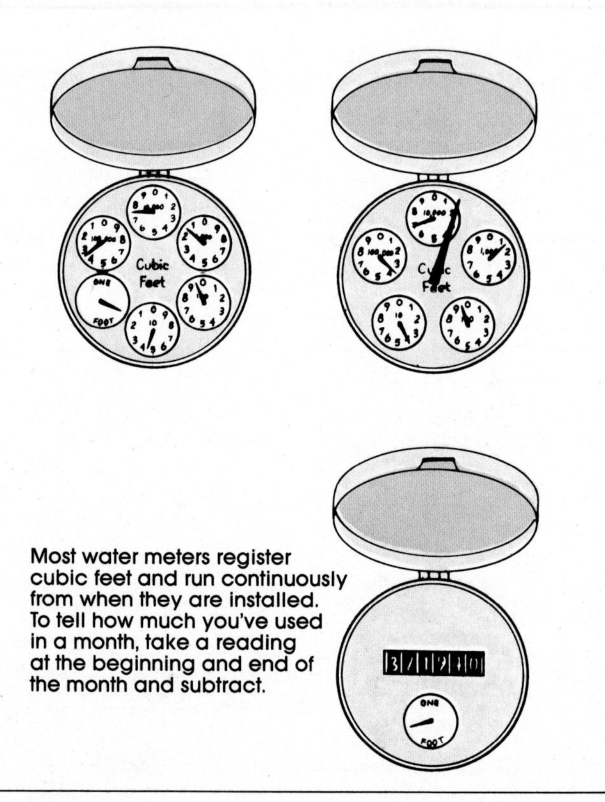

Most water meters register cubic feet and run continuously from when they are installed. To tell how much you've used in a month, take a reading at the beginning and end of the month and subtract.

on how much water you used. The water-use cost is usually an amount between 30 and 60 cents for each 100 cubic feet.

Multiply the cubic feet of water you used for each shower by the number of showers you take each month to get the cubic feet of water you use each month for showers. Then, divide this figure by 100 and multiply by the rate per 100 cubic feet.

Example:
Water used for one shower = 2.7 cu. ft. (20 gals.)
Number of showers per month = 36
Cost per 100 cu. ft. = $.55

Step 1. 2.7 × 36 = 97.2
Step 2. 97.2 ÷ 100 = .972
Step 3. .55 × .972 = .5346

The water for your showers costs you a little over 53 cents each month.

Unless you take only cold showers, the cost of water you use isn't the total cost of a shower. You must also pay the cost of heating the water. The gas company charges you for *therms*, not by the volume of gas you use. A therm is 100,000 Btu's (British thermal units). A Btu is the amount of heat it takes to raise one pound of water one degree Fahrenheit.

To figure the cost of heating your shower water, you multiply the number of gallons (36 × 20 = 720 in our example) by 8.3 (the number of pounds in a gallon) and this total by the number of degrees Fahrenheit you raise the temperature of the water. If the cold tap water temperature is 50 degrees and you like your shower water at 105 degrees, this number would be 55. This total is the number of Btu's you have used. Divide this by 100,000 to get therms and multiply the therms by the gas company rate per therm (let's say by $0.50).

Example:
Step 1. 720 × 8.3 × 55 = 328,680
Step 2. 328,680 ÷ 100,000 = 3.29
Step 3. 3.29 × 0.50 = 1.645

It costs a little over $1.64 to heat the water for your showers each month. Add this to the 53 cents for the water, and you find your showers cost you $2.17 per month.

Detecting hidden leaks. Another use for your water meter is to determine whether or not you have a hidden leak somewhere. A soggy spot in your garden, a wet place on a wall or ceiling, or an exceptionally large water bill can all indicate a leak. If you have suspicions but can't find an obvious leak, try this simple test. Turn off all the water everywhere in the house and garden. Watch the needles on your water meter carefully. If you see any of the needles moving, you have a pretty big leak. If none of the needles is obviously moving, use a felt-tipped pen to mark the exact location of the large needle. If there is no "large" needle, mark the small one labeled "one foot." Don't turn on the water for any purpose—or flush the toilet—for the next hour or two. Check the meter occasionally; if the needle moves in that period, you have a leak. To see how much water you are losing from the leak, use the same method described for finding out how much water you used for showers.

How to Check for Leaks

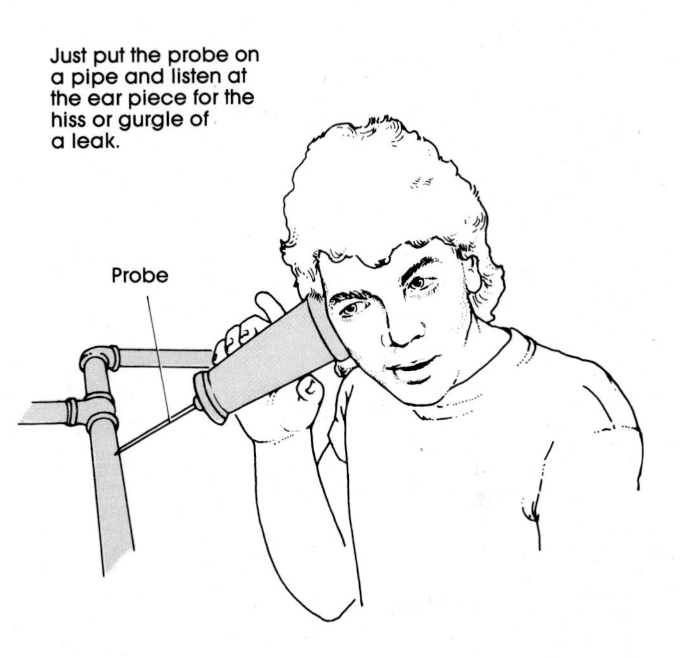

If the one-foot needle moves across three marks on the circle in an hour, you have used three-tenths of a cubic foot in that time. Your leak is wasting over 216 cubic feet per month—that's over 1,500 gallons, which is quite a few showers.

If you suspect a leak, but do not have a water meter to check it out as described, you can find it in other ways. Leaks tend to make noise. Sometimes it's a hiss and sometimes it's a gurgle. Put your ear against an exposed pipe, hold your breath, and concentrate on listening. The noise of a leak, unless it's quite a large one, can be very faint and difficult or impossible to hear without mechanical assistance. Plumbing supply stores carry listening devices that are designed to amplify the sound in pipes. These usually consist of an ice-pick-shaped pickup and an ear piece that resembles an old-fashioned telephone receiver. Just put the point of the pickup against the pipe and your ear against the ear piece. Usually the closer you are to the leak, the louder the sound tends to be—not always, but usually. So move around from place to place until the sound is loudest, and you will probably be pretty close to the leak.

Water companies are usually very cooperative in helping you find leaks. They will probably send an experienced person with a sonic leak detector to your house; such equipment can often isolate the area of the leak in a very short time. Don't hesitate to call if you think you are wasting water and money on a hidden leak. There is usually no charge for this service.

THE BASIC WATER SUPPLY SYSTEM

The labyrinth of pipes, fittings, valves, and fixtures that constitutes the plumbing system in your house can be a complex puzzle. However, like most devices and systems designed by human beings, it is only bewildering to those who are unfamiliar with it. If you are interested enough in the subject to read this book and look at the drawings, you will soon see that plumbing is just a series of rather simple and logical connections of pipe, fittings, and fixtures. You may find that making repairs or even adding on to the system is quite possible for you to do all by yourself.

Supply Lines and Branches

It all starts with your main supply line. If you get your water from a public water works, it will probably come from the water main under the street out front, through your water meter and your main shutoff valve. In areas where the ground freezes deeply in the winter, the meter will be in the basement or crawl space. In warmer areas, it will be in a little concrete well near the curb. If you have a private water supply system, the main supply line will start at your storage tank or pump.

In any case, the supply line that enters your house should be at least ¾-inch pipe. One-inch pipe is even better if you are planning any additions or replacements. Obviously, the larger the pipe, the more water it can supply. But even more important, a larger pipe will cause less friction. When the water pressure is low, like during TV commercials when everyone uses their bathrooms at the same time, both the volume and friction factors are important to maintain a steady, even flow of water.

After the main line enters your house, it usually runs near the water heater. There it splits in two. The hot-water pipe goes through the water heater and then runs parallel to the cold-water pipe to the laundry, kitchen, and bathrooms throughout the house. At one or more convenient places in the basement or crawl space, the cold-water pipe branches to provide water to the outside faucets around the house and in the garden.

If the water entering the house must be treated in some way, the treatment units will be attached near where the water enters the house. Chemical injectors or filters will be on the main line in the basement or crawl space. Because softened water usually doesn't taste as good as hard water and because the added sodium is not considered healthy, water softeners are often placed on the branch line that goes to the water heater. This way only the hot water, used for washing and bathing, is softened. In some cases, although it makes the piping a little more intricate, unsoftened water is piped to the kitchen sink and maybe a wet bar or faucet used primarily for drinking, while softened hot and cold water is piped to the laundry and bathrooms.

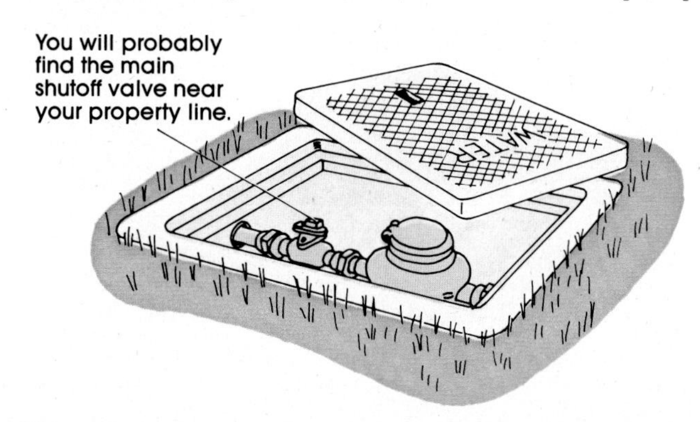

You will probably find the main shutoff valve near your property line.

To make the plumbing system more economical to install and to keep pipeline friction to a minimum, architects and house designers try to locate bathrooms, kitchens, and laundries as close together as possible on the same floors and, when they are on different floors, directly above and below one another. As you can see in the drawing on page 11, this makes for a much simpler installation and one that uses much less pipe and fewer fittings.

After the main supply line enters your house, it divides into ¾-inch hot- and cold-water pipes. The branches that feed individual fixtures may be of ½-inch pipe. These water systems may be made of galvanized iron, copper, or plastic, depending on local codes and practices and the age of the house. Building codes are very specific as to what kinds of pipe and fittings are required or allowed. Before you make any changes or additions, especially if you plan on changing from the kind of pipe already in your house, be sure to check with your building inspector.

In cold climates where pipes can freeze—and sometimes in warmer areas, too—the entire supply system is sloped to low points where drain cocks will allow the drainage of all the pipes in the system. If your system is set up this way and you are adding to it, be sure to slope your new pipes slightly toward the drain cock.

Each fixture and appliance, without exception, should have its own shutoff valves. Water softeners, filters, and other treatment devices, water heaters, dishwashers, washing machines, as well as all the sinks, tubs, and toilets should have a valve on both the hot- and cold-water pipes. If you plan on doing some or all of your own plumbing repairs and maybe some additions, installing any missing shutoff valves would be a good place to start and give you helpful practice.

Another constituent of a properly installed plumbing system is the air chamber or cushion. Air or inert gas trapped in a chamber compresses and cushions the shock when a nearby faucet or valve is turned off suddenly. Because water is not compressible it makes a loud bang and can damage pipes when it is stopped suddenly without this kind of cushion.

An air chamber is a vertical section of pipe on a supply line with a cap on top. It is usually hidden beneath cabinets or within the wall. The air in a pipe air chamber eventually dissolves in the water or leaks out around the cap, making it useless as a cushion. The cap must then be removed and a new supply of air allowed to enter.

If the cushion must be put within the wall or in another location where it is not easily serviced, consider installing a manufactured shock absorber. These are made with inert gas and a bellows so they remain permanently effective without servicing.

Water Supply and Drain Systems

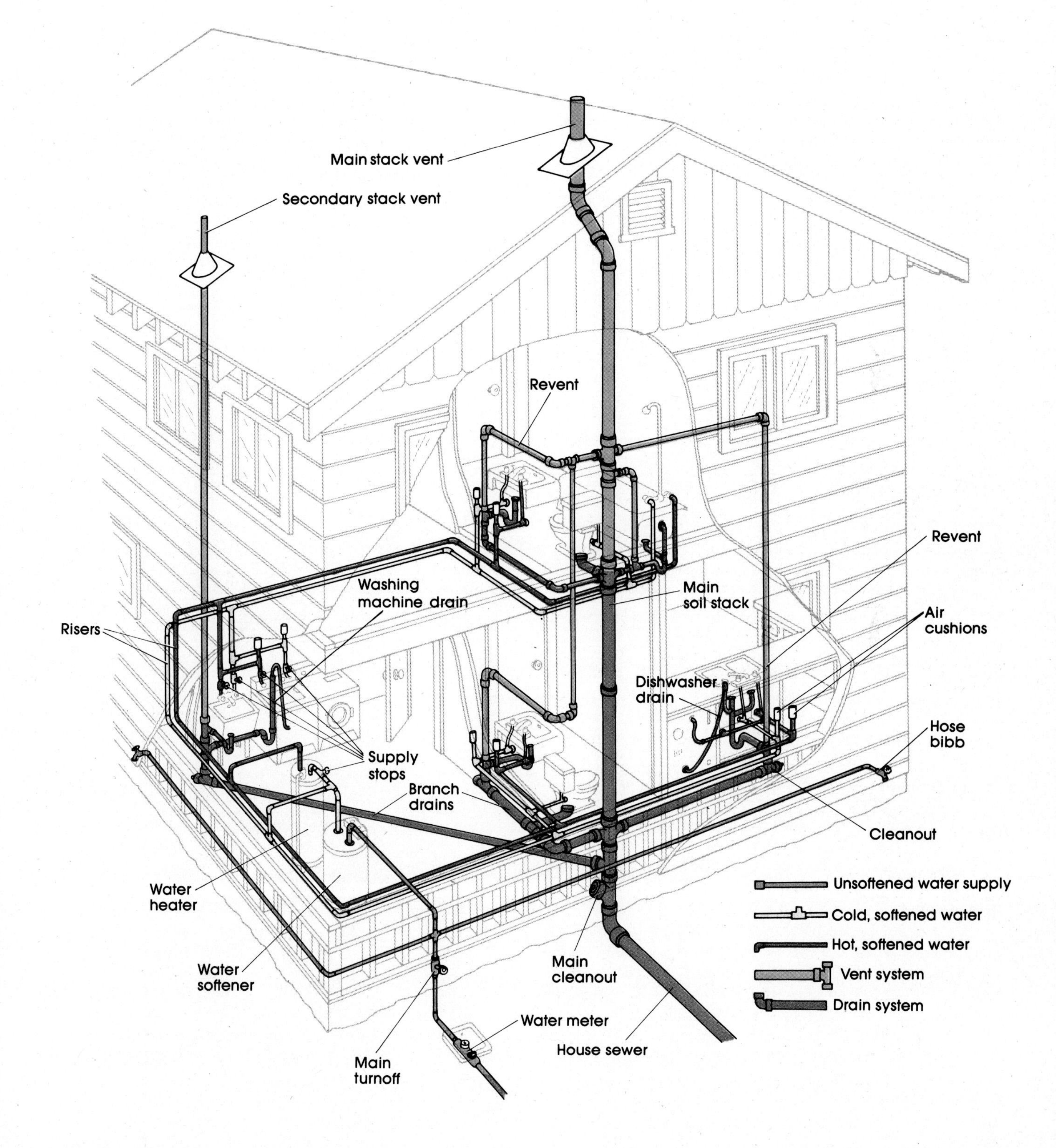

The Hot-Water System

A home hot-water system consists of a heater and a piping system that parallels the cold-water pipes to the faucets where hot water is desired. The heater is fueled by gas, oil, electricity, or the sun, depending on the fuel that is available in the area and the preference of the homeowner. Most heaters cannot heat water as fast as it can flow from a shower head or faucet, so they have a tank in which to store a quantity of hot water.

Home water heaters are generally available with tanks of 30- to 82-gallon capacity. The once-common 20-gallon tanks have become so unpopular that most manufacturers don't make them any more. Gas and oil-fired heaters are usually 30 to 60 gallons. Electric heaters, because they heat water more slowly and have a longer recovery time, have tanks that can hold up to 82 gallons. Some manufacturers make a *rapid-recovery* electric water heater in which the upper heating element operates independently to heat the top quarter of the tank quickly. When the top quarter is hot enough, the upper element goes off, and the lower one comes on to heat the rest of the water slowly.

The size of your water heater tank depends on how much hot water your family needs at the time of peak usage—for example, in the morning or evening when everyone takes a bath or shower. Usually the number of bedrooms in a house is used to determine the size of the water heater. For a one- or two-bedroom house, a 30-gallon tank is recommended; for three bedrooms, a 40-gallon tank, and so on. If you have several children, you may want to have a larger tank than recommended.

In some areas of the United States *demand* water heaters are available. Also called *tankless* or *instantaneous* water heaters, they are common in Japan and Europe but haven't been used in this country to any great extent since the copper shortage during World War II. These heaters have an intricate grid of copper ducts very much like an automobile radiator. When the hot water is turned on and flows through the heater, a large gas flame envelops the grid and heats the water to the desired temperature as you use it. Because this kind of heater heats water only as you use it and does not have to go on periodically to keep a whole tank of water hot, it uses up to 20 percent less fuel than a conventional tank heater. You can probably find out about availability in your area through a wholesale plumbing supply or solar equipment dealer.

Tank-type water heaters are basically simple devices. They consist of a thermostatically controlled burner or heating element that heats the water and an insulated tank to hold it until someone wants to use it. Because hot water is corrosive to metals, most quality heater tanks are glass-lined to help prevent them from rusting through.

Gas-fueled heaters have a burner very similar to the one on a gas kitchen range. A pilot light lights the burner when the thermostat indicates the tank water is cooler than desired. Hot exhaust gases from the burner go through the flue in the center of the tank, where they continue to heat the water as they pass by.

Oil-fueled heaters are less common but still in use where oil-fired furnaces are used. The water tank is suspended in the middle of the heater, and a small version of an oil-furnace blower heats the bottom and sides with hot gases before they are vented out the flue.

Electric water heaters have one or two heating elements that project through the wall of the tank into the water. Because there is no combustion, no flue is needed and no toxic gases or hot gases are emitted. An electric

Types of Water Heaters

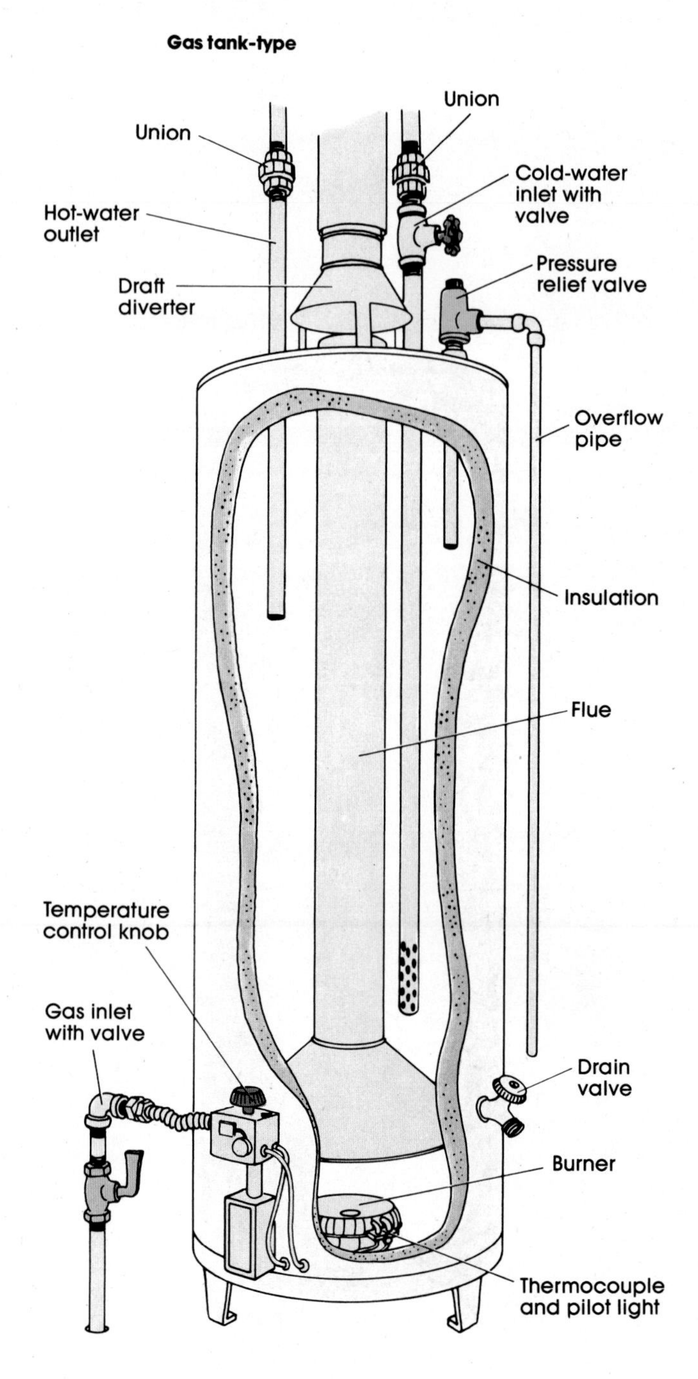

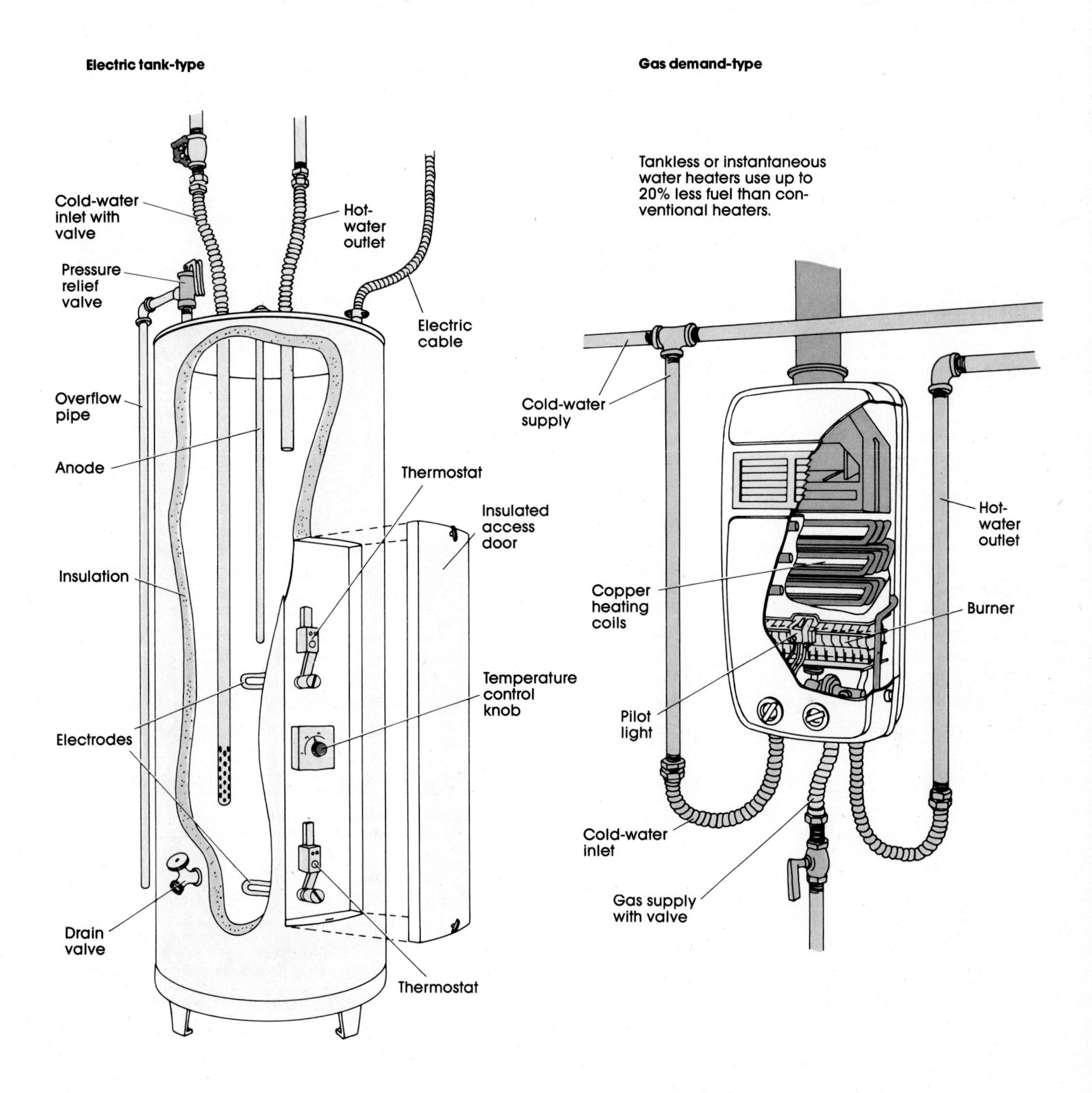

heater, therefore, can be placed in a closet or bathroom, where the building code would not allow a gas or oil-fueled heater. All water heaters do need a drain in case of overflow and for regular maintenance draining (see page 76).

All water heater storage tanks should have a temperature/pressure relief valve. This valve will relieve the pressure and thereby prevent an explosion if the thermostat malfunctions and the temperature or pressure within the tank exceeds predetermined limits. A properly installed relief valve will have an overflow pipe attached to it that directs any escaping water or steam outside or to a floor drain.

For information on passive, active, and auxiliary solar water heating systems, and what might be the best system for you, see pages 78–79.

WASTEWATER DISPOSAL

At the end of the water supply pipes in your home are fixtures or appliances where the water is used. The system that carries the used water away is the wastewater drainage system, or, as it is called in the trade, the DWV or drain-waste-vent system. Because of the nature of all the household and body wastes that are put into our used water, we need more than just drain pipes to take this water away. Our waste materials contain large amounts of bacteria, both beneficial and harmful varieties. The anaerobic bacteria, those which live in an environment without free oxygen, thrive in our digestive tracts. When these bacteria are in septic tanks, and sewer lines, they produce a foul-smelling, poisonous, and flammable mixture of gases, commonly called sewer gas. This gas must be kept out of our homes.

To prevent sewer gas, pests such as rats, and other contaminants from coming up the drain pipes into our homes, we add traps to the system. A trap is a U-shaped curve in a pipe or fitting that remains filled with water at all times. Sewer gas and other things cannot get past the water and come up through the drains. Each time you empty a sink or flush a toilet, the water in the trap is also flushed away and new water replaces it. This means, theoretically, that dirty water isn't in the trap long enough to grow large colonies of its own bacteria.

To get rid of the sewer gas and to prevent siphoning of water out of the traps, all house drain systems have vents. Vent pipes come off the drains downstream from the traps and go up through the roof. Sewer gas passes up the vent pipes and is dispersed harmlessly into the air. When water flows down the drain, air is sucked down the vents into the pipes, equalizing the air pressure on each side of the trap. This prevents the water in the trap from being siphoned out and sewer gas from entering the house.

Also, since all drain systems are subject to clogging from time to time, they are provided with cleanouts. These are Y or T fittings with screw-on covers that give the homeowner or plumber a place to insert rods or augers for the purpose of dislodging clogs.

Drain pipes. Drain pipes differ from water supply pipes in at least two important ways. They are larger and they flow at a lower pressure, usually by gravity alone. The smallest drain pipe in the house, probably the one coming from the bathroom sink, is never smaller than 1¼-inch and 1½-inch is usually preferred by the building code. The smallest drain from a toilet is 3 inches in diameter. Metal drain pipe 2 inches or less in diameter is usually made of galvanized wrought iron, galvanized steel, or copper tubing. In larger sizes, the pipe is made of cast iron. Most building codes now allow plastic pipe for DWV systems in single-family homes, but there are usually some restrictions (see page 38).

The DWV system in a house is arranged like a tree, with the smallest branches near the top. The smallest pipes always flow into pipes the same size or larger until they all flow into the soil stack, building drain, and building sewer, which are the largest pipes in the system. The sizes of all these pipes are specified in the building code. A typical arrangement of pipes with their usual sizes and names is shown in the drawing.

If you are going to replace a section of pipe or replace an old fixture with a new one, use the same sizes of pipe, trap, and vent that were there before. If you are putting on an addition that includes a new bathroom or laundry, you should check with the code to be sure of the required pipe sizes. Unlike some specifications in the building code, pipe sizes are not minimum sizes. The theory of pipe sizing prescribes sizes that are neither too small nor too large, but just right. Pipe that is too small won't allow the proper amount of material through and may clog or unduly retard the discharge of wastewater.

Not so obviously, horizontal drain pipe that is too large is also undesirable. Larger pipe results in slower movement of the waste material through the pipe. This slower speed reduces scouring and lets heavier particles settle. The greater thermal mass of the larger pipe also causes greater cooling and solidifying of grease. All these

Detail of Part of the DWV System

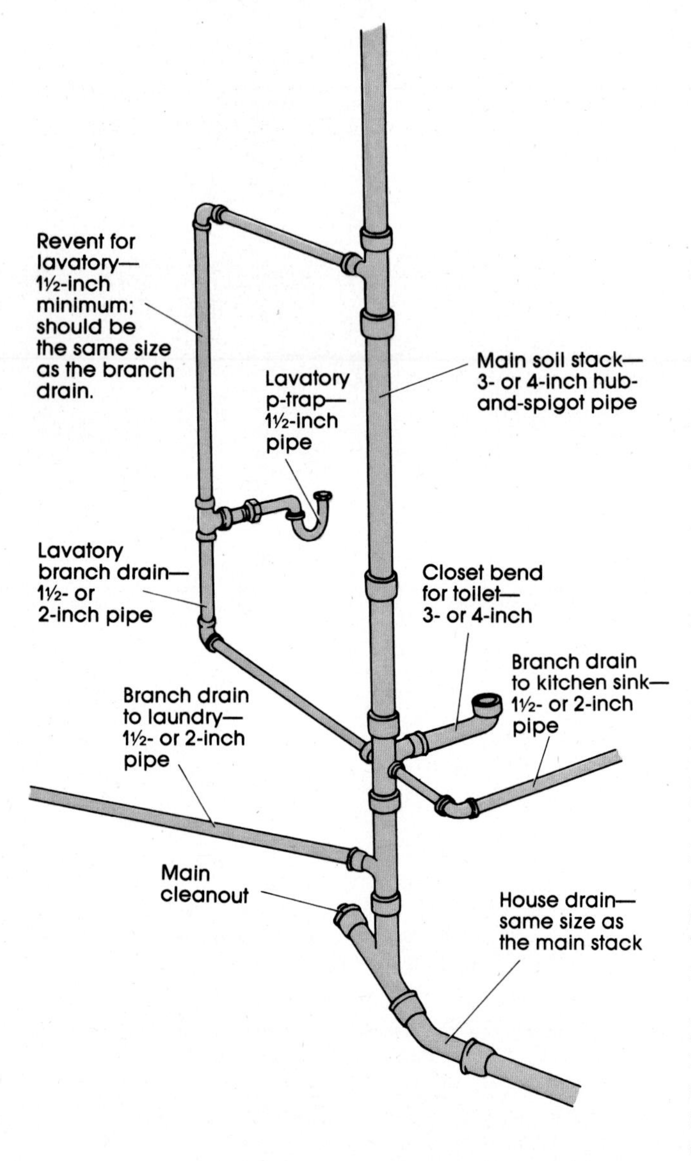

things contribute to the possibility of clogging. Unnecessarily large pipe also adds unnecessarily to the cost, whether the pipe is horizontal or vertical.

The proper size pipe allows for enough capacity to prevent back-ups, to promote the scouring action of swift movement, and to avoid siphoning or blow-back by allowing enough vent air. For drain pipe it is recommended that you always use the smallest size permitted by the code.

Besides making the drain pipe smaller, there is another way to increase the velocity of the waste within the pipe. That is to increase the pitch or slope of the horizontal parts of the system. Codes require that horizontal branches slope ¼ inch per foot. In practical application, this slope varies from 1/16 to ½ inch per foot, and of course the more the pitch, the faster the water will flow. If you increase the slope of a fixture drain pipe, be careful not to make the outlet from an unvented length of pipe lower than the bottom of the fixture trap it serves. If it is lower, it is likely to siphon the water from the trap (see drawing).

Vent system. Building codes require a vent system that is arranged and sized to provide the best possible pressure/suction relief for each fixture in the system. Systems that don't follow the code may have back pressure problems such as sinks that drain too slowly, toilets that need several flushes to get rid of all their contents, and blow-back through first floor fixture traps. A poor vent system can also provide too much "negative pressure" (suction), which siphons the water from traps, or "positive pressure", which forces bubbles of sewer gas through the liquid in the traps (blow-back). Either way the smell is not desirable.

The size of vent pipe cannot be smaller than required by the code. The sizes are based on the kind of fixture, the diameter of the drain being vented, and the length of the vent pipe. For each dwelling unit, most codes require a 3-inch or 4-inch main stack extending through and above the roof. The diameter of an individual vent can never be less than 1¼ inches or less than one-half the diameter of the drain it serves, whichever is larger.

Traps. Building codes everywhere require that each fixture connected to the household drain system have a water-seal trap. Some fixtures, toilets for example, have their trap built in as an integral part of the fixture. The trap most commonly used in the home is called a P-trap because it looks like the letter "P." It is used on sinks, laundry trays, and most other fixtures that don't have built-in traps.

Building codes have many restrictions on traps because they are so important in protecting the health and welfare of the home's inhabitants. The following restrictions appear in virtually every code:

■ Traps must be self-cleaning. That is, they must be smooth inside so hair, lint, and other material cannot be caught and retained.

■ No trap can depend on moving parts for its seal.

■ No trap outlet can be larger than the fixture drain it is serving.

■ Each trap must have a water seal no less than 2 inches or more than 4 inches.

Trap Warning

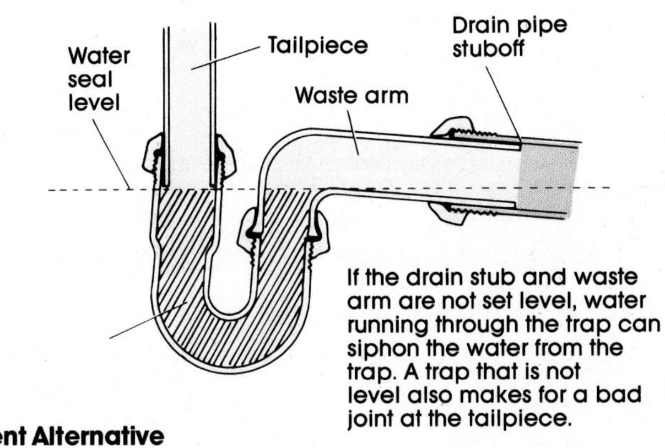

If the drain stub and waste arm are not set level, water running through the trap can siphon the water from the trap. A trap that is not level also makes for a bad joint at the tailpiece.

Vent Alternative

Instead of a secondary stack or a revent for a sink or tub, you can use an automatic vent valve. It cannot be sealed in a wall but must be accessible for inspection.

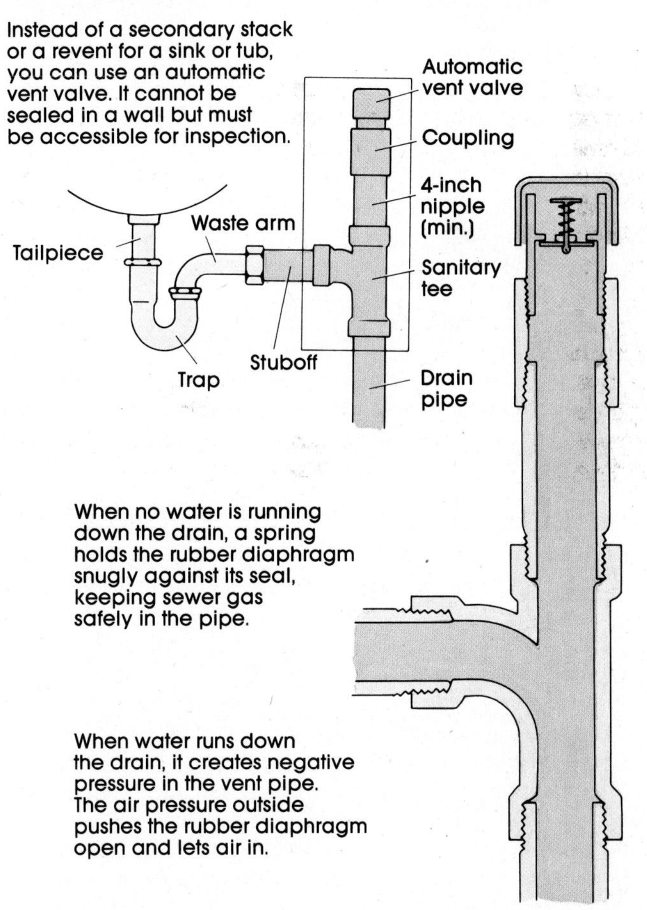

When no water is running down the drain, a spring holds the rubber diaphragm snugly against its seal, keeping sewer gas safely in the pipe.

When water runs down the drain, it creates negative pressure in the vent pipe. The air pressure outside pushes the rubber diaphragm open and lets air in.

■ All traps must be installed level in relation to their water seals to prevent siphoning.

■ Each plumbing fixture must have its own trap. There are some exceptions whereby installations of sinks and laundry trays, when there are two or three units adjacent to each other, can be connected to a single trap.

■ No fixture can be double trapped. A toilet with an integral trap, for instance, cannot be connected to another trap.

Typical P-Trap Assembly

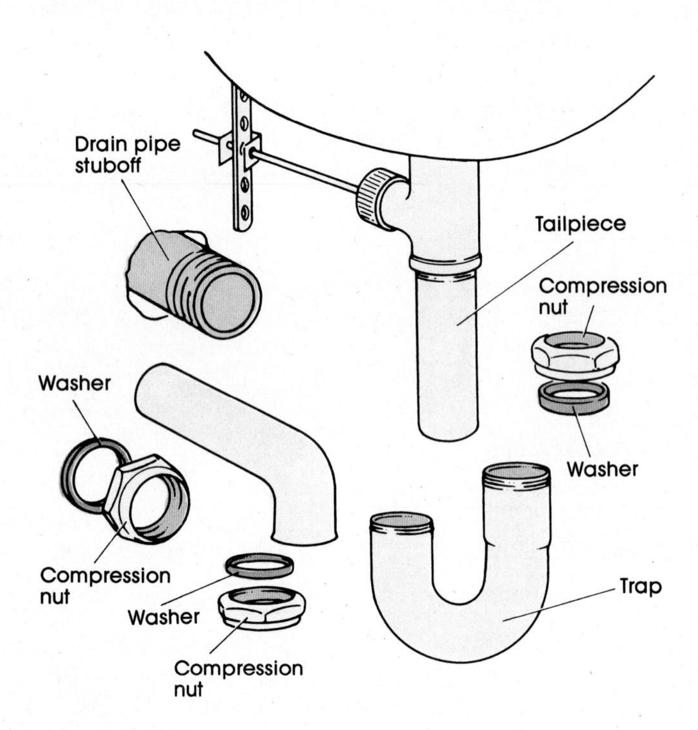

Again, if you are replacing a fixture or adding fixtures similar to those already in your house, use a trap like those already in use and you will most likely be well within the requirements of the code. If you are installing a new fixture unlike anything you already have, consult a plumber or your building inspector for advice on the local requirements.

Cleanouts. All building codes now recognize the importance of accessible cleanouts and require them on all installations. The requirements regarding location, size, and minimum distance between cleanouts are spelled out very specifically. For example:

- A cleanout is required where the building sewer crosses the property line or connects to the public sewer. This cleanout allows for the cleaning of stoppages that may occur in the public sewer lateral and also is the place where tests can be performed on the entire house system.
- Some codes require and some allow a cleanout outside the house within five feet of the building.
- Accessible cleanouts are required on all horizontal drain lines and where there is a change in direction of more than 45 degreees in the building drain.
- A cleanout is required at the base of all stacks.
- All cleanouts must have at least 18 inches of clearance to allow the access of cleaning rods, snakes, and other tools.

Sometimes in one-story buildings, the local code will consider a roof vent to be a cleanout for the stack it serves. Certain requirements involving pipe sizes and changes in direction must be met for this to be the case, however. Be sure to check your local code.

Drainage from Below Sewer Level

Sometimes it is necessary to dispose of wastewater that accumulates below the main sewer line. Homeowners will occasionally want to install wash trays, an automatic washing machine, or even a whole bathroom including a sink and a toilet in a basement area. When groundwater seepage or rainwater runoff repeatedly floods a basement, facilities must be installed to automatically remove the water. Discharge of rainwater into a sanitary sewer is not permitted in some areas.

Sump pumps. Groundwater, rainwater, or gray water—water from a washing machine, bathroom sink, shower, or bathtub—can be handled very well with a sump pump. The wastewater is made to run into a concrete-lined sump pit and is then lifted into the sewer with a pump that starts automatically when water reaches a certain level in the pit.

If seepage or rainwater is flooding the basement, the sump pit should be located at the lowest point in the basement. A sump pit for a sink or washer can be anywhere that it's convenient to run a drain pipe. The size of the pit will be determined by the size and kind of pump you install.

The oldest, most common, and least expensive sump pump is the upright type. It consists of an electric motor on top of a pedestal. The base of the pedestal, containing the pump and discharge pipe, rests on the bottom of the sump pit. A ball float in the pit is connected by a rod to the motor switch. When the water in the pit reaches a predetermined level, the float flips the switch to ON. When the pump lowers the water level, the ball descends and turns the switch to OFF. The water level to turn the switch off is usually set at about 6 inches because the pump can be damaged if it empties the pit and runs dry. The pit for an upright sump pump should be 12 to 24 inches deep and 12 inches or more across; it can be either round or square.

The submersible pump is more expensive, but generally more satisfactory than the upright type. The submersible pump cannot be damaged by flooding and requires much less maintenance. It can run longer and safely take the water level lower than the upright pump. Two kinds of switch mechanisms are available on these pumps. One is the float type similar to the one described on the upright pump. These are subject to jamming if dirt or other debris accumulates in the sump pit. The more desirable type of switch is activated by water pressure. This one, too, can be set to go on and off at specific depths. The sump pit for a submersible pump should also be at least 12 inches across and from 12 to 15 inches deep.

The installation of a sump pump is a relatively simple project. The first step is to prepare the pit, which should have a level concrete bottom and sides of either concrete or a section of concrete or terra cotta pipe 12 inches or more in diameter.

Break a hole in the basement floor and dig a hole a little wider and 3 or 4 inches deeper than the pit will be. Put 3 or 4 inches of gravel in the bottom and set a wooden form or pipe section into the hole (see drawing).

Pour concrete in the bottom of the hole and around the pipe or form. Smooth and level the bottom and smooth the concrete around the edge even with the basement floor.

When the concrete is set, put the pump in place and make all the necessary pipe connections leading into the sewer line or seepage pit. A check valve (see page 50) and union (see page 27) are desirable, in fact mandatory in many codes. Before you plug in the pump, be absolutely sure that the receptacle is grounded.

You can make a cover for the sump pit of ¾-inch exterior plywood. For the upright pump, three holes will have to pierce the cover—one for the pedestal, one for the float ball rod, and one for the discharge pipe. For the submersible pump, only a hole for the discharge pipe will be needed. Drill the hole(s) and then cut the cover in half through the hole(s). Put the pieces together over the pit and hold them together with straps of aluminum or wood held with screws.

Up-flushing toilets. In some parts of the United States, mostly in the Northeast, single-unit up-flushing toilets are available and allowed by building and health codes. These toilets usually depend on high water pressure, 40 psi or more, to break up the solids into small pieces and siphon them upward to an overhead drain line. Up-flushing toilets are fairly easy to install yourself. They usually come with complete, illustrated instructions that must be followed exactly.

Another alternative that is more universally available is a sewage ejector, which is a tank with a pump inside. Material from a regular toilet is flushed into the tank and then pumped up, as much as ten feet or so, to the sewer line. You may find that it's more convenient to walk upstairs to go to the bathroom than to pay for this installation, however. The cost of a toilet, the ejector, and all the pipe and fittings will run you around $700 or so, even if you do all the work yourself. Having someone install it could increase the cost to $1,000 or more.

The sewage ejector tank is about 30 inches high and 20 inches in diameter. A 4-inch inlet pipe enters the side of the tank about 20 inches below the top, and a 4-inch discharge pipe comes out the top to be connected to the sewer line overhead. Since the ejector tank inlet must be below the toilet discharge, there is no way of installing one of these without excavating below the basement floor (see drawing).

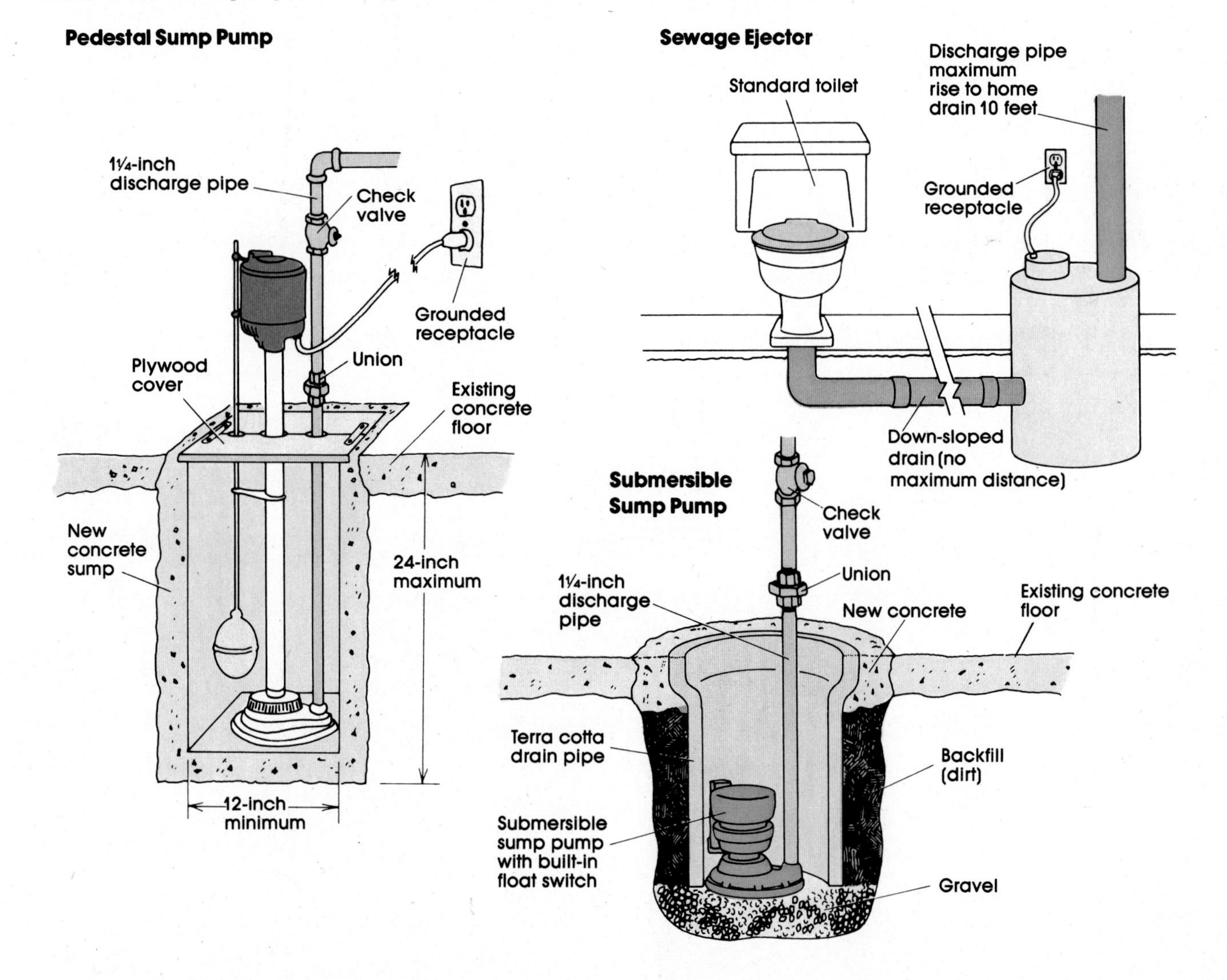

EMERGENCY REPAIRS

What to do for the problems that can't wait—an overflowing toilet, clogged drain, frozen pipes, overflowing washer, or leak in a pipe or water heater. Gather the tools you need to keep handy for any plumbing emergency.

Before an emergency occurs, you can do a few things to minimize the damage and make any problem easier to handle. Learn where all the water and gas shutoff valves are. Have a basic set of plumbing tools all together in a handy place (see page 6). And keep a stock of a few parts and materials just in case of a plumbing emergency at night or on a holiday—when most emergencies seem to happen.

If you have a water meter, its shutoff valve can be locked by the water company. This valve may take a special tool to operate and have holes for a padlock so if locked there's no way you can turn it back. In snow areas, the meter will be in the basement rather than in a meter box.

Your main water valve, sometimes called the main *supply stop,* should be somewhere between the water meter and your house. Usually it's either near the meter or near where the main water line enters the house. In cold-weather areas, the meter may be within the basement or crawl space, and you will probably find the main valve very close to the meter.

If the plumbing in your house has been designed and installed properly, you may only have to use the main water valve on rare occasions. Each fixture or appliance that uses water should have a supply stop for each pipe carrying water to it: usually one for the cold water and one for the hot. You will find these valves under sinks and toilet tanks, on top of the water heater and behind washing machines and dishwashers. Many codes require separate stops for each fixture, and you rarely find them on showers or tubs. If you cannot find shutoff valves near all your fixtures and appliances, installing them might be a good way for you to get started as a do-it-yourself plumber (see page 54). Some plumbing systems have additional shutoff valves where branch lines leave the main line to serve the various parts of the house.

Find all your supply stops. If it's not obvious what part of the system is served by each valve, turn one off and check what doesn't work. Then label the valve so you won't waste time in an emergency. You always turn stop valves clockwise to shut them off.

◀ This chapter will help prepare you to meet plumbing emergencies, large and small.

Emergency Plumbing Kit

In addition to your basic set of tools, here are a few things to keep handy:

- Some sheets of old rubber, like part of a tire, an inner tube, a water bottle, or kitchen gloves
- A length of old garden hose or radiator hose
- A piece or two of sheet metal—an old coffee can will work just fine
- A roll of duct tape or plastic electrical tape
- A few nuts, bolts, and washers of various sizes
- A couple of automotive hose clamps
- Assorted faucet washers and screws—some hardware stores have prepackaged kits that contain various rubber, fiber, and packing washers and screws
- A few assorted O-rings
- A couple of wire coat hangers
- Pipe joint compound

For anxiety-free maintenance of your faucets, put together a kit of assorted washers, O-rings, screws, and packing string.

CLOGGED TOILETS & DRAINS

If the water persists in running in your toilet tank, remove the top and check out why. The tank ball or flapper may be stuck open. If so, free or adjust the rods or chain holding it open so it can close. If the tank ball or flapper is worn so it won't close tightly, it will have to be replaced (see page 61). If the water level is too high so water is going into the overflow pipe, pull up on the float ball. If this shuts off the water, your problem is a ball that is set too high. Some mechanisms have an adjusting screw at the base of the float ball rod. If yours does, turn the screw to move the ball lower in the tank. If there is no adjustment screw, bend the rod to lower the ball. The ball should be adjusted so the water level is ½ to 1 inch below the top of the overflow tube.

If the bowl fills and won't drain or, worse yet, overflows all over the bathroom, use a plunger quickly. The kind of plunger with a cone works best on a toilet (see page 7). Press the plunger into the drain hole and work it up and down rapidly a dozen times or more. If it won't drain right away, keep the bowl at least half full of water and pump as much as you can for 10 or 15 seconds. If it still won't drain, rest for an hour or so to let the clog soften and then try again.

If the plunger treatment fails, use a closet auger, which has a bent end designed to protect the porcelain. These can be rented, but buying one is much cheaper than even one visit by a plumber. Feed the bent end of the tube into one drain hole and then feed the auger through it into the drain and turn the crank clockwise until you reach the obstruction. Push a little more; then pull the auger out while continuing to turn the crank clockwise. If you can't dislodge the clog with a plunger or reach it with an auger, the obstruction may be down in the main drain (see page 21).

Unclogging a Toilet

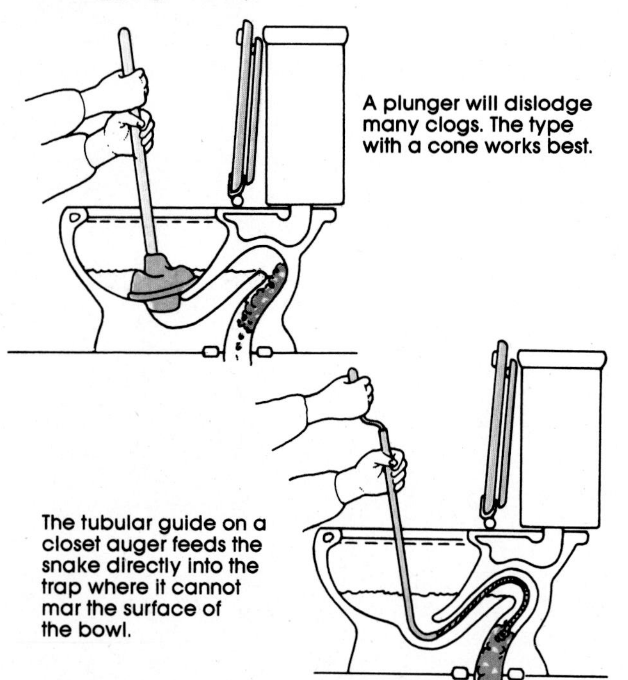

Sinks and Lavatories

Do not use chemical drain cleaners. They are caustic and could burn you if you had to drain or work on the pipes. Chemical cleaners can also form a permanent clog by melting the obstruction into one solid piece before the water can wash it away.

The stoppage may be in the P-trap just below the sink or it may be farther down the line. Start by removing the sink strainer and cleaning it. It can be removed by prying or by removing two small screws. If a stopper is in the way, it can usually be removed by giving it a quarter turn and lifting.

Use a piece of coat hanger wire with a small hook on the end or a snake to reach down into the P-trap and see what you can hook onto. If nothing seems to be there, follow the steps shown on page 21. If you still can't reach a clog, it's probably in the main drain pipe (see page 21).

Bathtubs

Sometimes a tub with a built-in drain plug doesn't drain properly because the adjusting nut on the drain plug linkage has worked loose, and the drain doesn't open wide enough. To adjust this, unscrew the drain and pull the linkage out. Tighten the nut on the threaded screw to push the plug farther out of the drain when it's open.

If your tub doesn't have a built-in plug or the plug isn't a problem, then the drain pipe must be clogged. The first step is the same as with a sink—use your plunger, following the explanation on page 21.

In an older house or apartment building, you may have a drum trap in the floor near the tub instead of a P-trap beneath the tub. If you do, remove the top from the trap with a monkey or crescent wrench. Run your snake first through the lower pipe back toward the tub. If there's no clog there, run the auger down the upper pipe toward the main drain. Again, if you can't reach a clog, it's probably in the main drain.

Showers

A shower drain can be difficult to clear with a plunger, but it's worth a try. Coat the bottom of the plunger cap with petroleum jelly for more suction and keep an inch or two of water in the shower pan. If you seem to be getting good suction, keep up the plunging for several seconds as with sinks and tubs. If the plunger doesn't seem to be working, go get your garden hose.

The next step is shooting water down the drain with a garden hose to try to dislodge the clog. For this attempt, you'll need a helper. Attach your hose to a hose bib and bring the other end to the shower through the bathroom window. If you can't reach a hose bib with the other end of the hose, buy an adaptor at the hardware store so you can hook the hose to the bathroom sink faucet.

Remove the strainer from the shower pan and shove the hose a foot or so down the drain. Stuff rags around the hose and stand on them to seal the drain and to prevent the hose from jetting itself out of the drain when you turn the water on. You probably won't be able to reach the faucet handle while you're standing in the shower so have your helper turn the water on and off

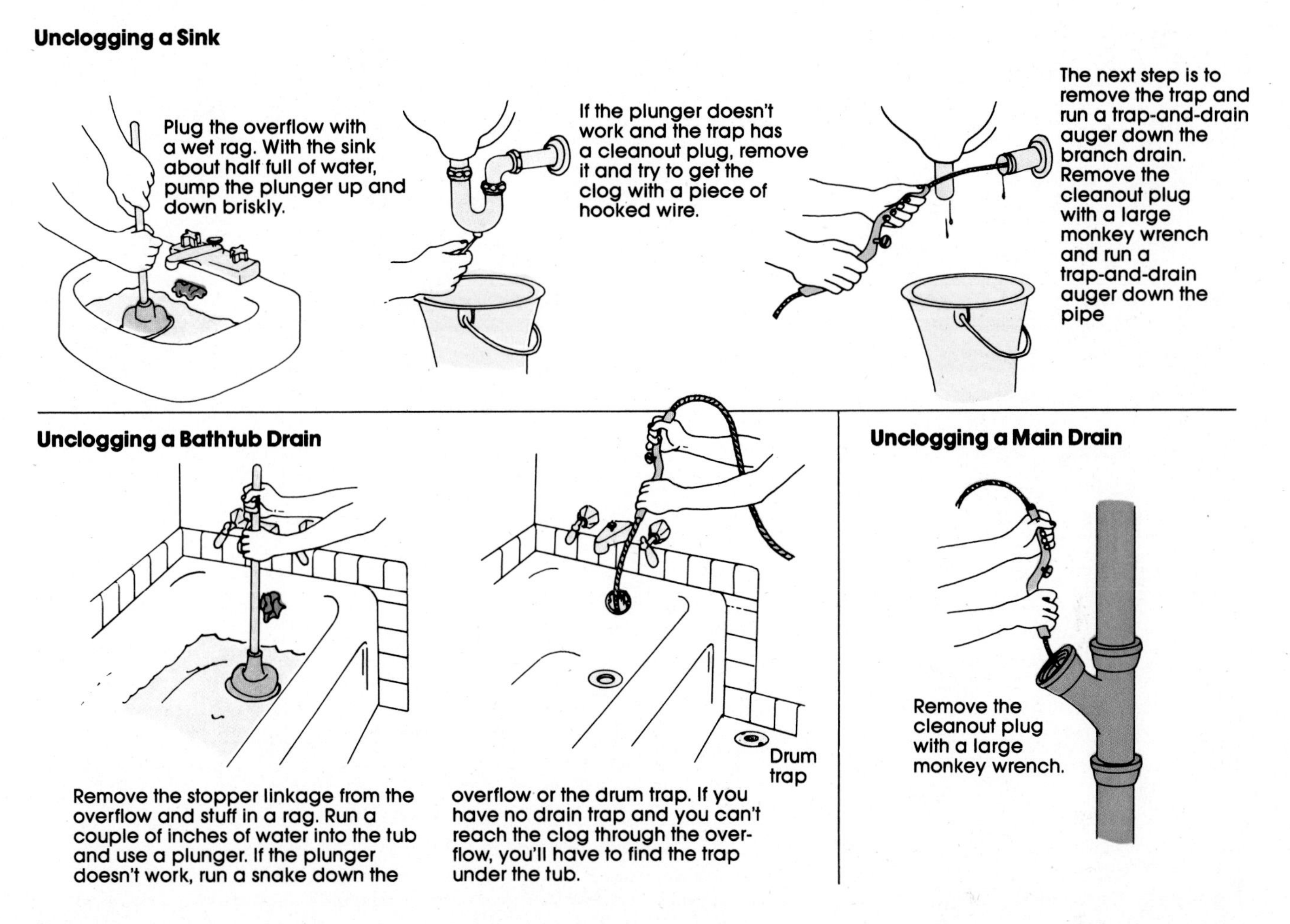

Unclogging a Sink

Plug the overflow with a wet rag. With the sink about half full of water, pump the plunger up and down briskly.

If the plunger doesn't work and the trap has a cleanout plug, remove it and try to get the clog with a piece of hooked wire.

The next step is to remove the trap and run a trap-and-drain auger down the branch drain. Remove the cleanout plug with a large monkey wrench and run a trap-and-drain auger down the pipe

Unclogging a Bathtub Drain

Remove the stopper linkage from the overflow and stuff in a rag. Run a couple of inches of water into the tub and use a plunger. If the plunger doesn't work, run a snake down the overflow or the drum trap. If you have no drain trap and you can't reach the clog through the overflow, you'll have to find the trap under the tub.

Unclogging a Main Drain

Remove the cleanout plug with a large monkey wrench.

abruptly several times. Use only short bursts of water pressure here. Gravity drains are built to withstand little or no pressure and some joints could be damaged by the pressure of the supply line.

There is a specially made and relatively inexpensive nozzle for this chore that you can probably buy at your hardware store or plumbing supply dealer. It inflates when the water is on to seal the drain and at the same time shoots a narrow, high-pressure stream of water down the pipe. This nozzle is made by several manufacturers, but there doesn't seem to be a common generic name for it. You'll just have to describe it.

Main Drains

If several fixtures are not draining properly or if you can't reach the blockage through a fixture drain, the blockage is probably in a main drainpipe. Or if a sewer odor begins to invade your house, there may be a clog in one of the vent pipes.

The easiest way to reach such blockages is through the vent terminals on the roof. Check to see that your trap-and-drain auger is long enough to reach to the bottom of the main stack from the roof. If you live in a two-story house, it probably isn't. In that case you may want to rent a 50-foot power auger from your local rent-a-tool store.

Once on the roof, be sure you position yourself very securely. Use a hooked ladder on a safety rope if you don't have a flat roof. Work the auger down the vent, cranking clockwise until you engage the clog. Keep cranking in the same direction as you try to pull it out or dislodge it.

If working down from the roof doesn't clear the blockage, you'll have to work from under the house or in the basement. You'll find one or more cleanout fittings with plugs, one at the bottom of the main stack and maybe one or two others to clean the large branch lines. You'll need a large crescent or monkey wrench to remove the cleanout plugs.

Before you remove the cover, remember that there may be several feet of wastewater backed up in the pipe. If possible, run no water in the house for several hours before you try this. By then most of the water may have seeped through the clog. In any case, have several buckets, pans, mops, and rags ready to clean up the mess when you remove the cover.

Work the auger into the Y-fitting and down the drain toward the sewer. Here again you may need to rent a long, powered auger to do the job. Once the obstruction is cleared, flush the pipe with clean water from a garden hose. When you replace the cleanout cover, coat the threads with pipe joint compound.

Some houses have a U-shaped house trap, which you can recognize by the two adjacent cleanout plugs.

This trap is usually located where the main drain line leaves the house. Always unscrew the plug closest to the sewer first. Do it slowly with rags and mop handy. If little or no water runs out, the obstruction is probably toward the house. If a lot of water comes out, the clog is down the main line toward the sewer.

If the clog is toward the house, work your auger slowly back up the pipe. To avoid a big mess, when you hit something, try to open just a small hole. Then quickly recap the trap and let the water drain through before uncapping the trap again and removing the clog.

A clog toward the sewer could be the result of tree roots in the pipe. You may have to rent a powered auger with a root cutting attachment or get professional help.

Frozen Pipes

Of course, if you live in an area where it gets cold every winter, your pipes should be protected already and you shouldn't have this problem. However, if they aren't protected, here are a few precautions to take in case an unexpected cold snap arrives.

Keep a pencil-sized trickle of water running from each faucet during the night. Moving water freezes much more slowly than water that's standing still. If possible, keep a light bulb burning near exposed pipes or put a portable heater in your basement for the duration of the cold spell.

For more permanent protection, wrap all exposed pipes in special pipe insulation. It usually comes with tape to keep it tightly and neatly in place. For even more protection, wrap your pipe with specially designed heating wires. They come equipped with thermostats so the heat is turned on only when it's needed.

Thawing Frozen Pipes

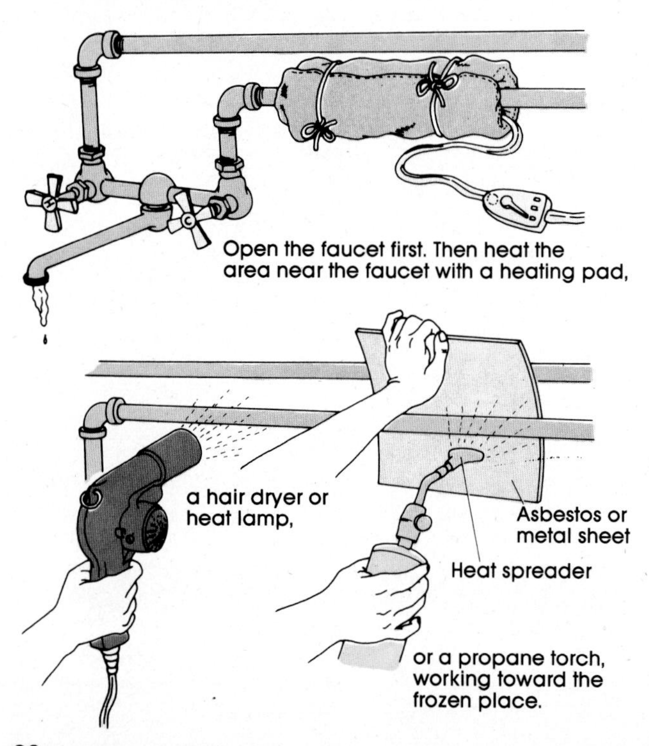

If it's too late for prevention—the pipes have already frozen—here are some ways to thaw them out.

Open a faucet near the frozen place in the pipe before you do anything else. If you're not sure exactly where the pipe is frozen, open all the faucets that are blocked. When heat is applied, vapor from the melting ice needs a place to escape, and if there is no opening through a faucet, the vapor may come through the side of the pipe. As you thaw a pipe, work back from the open faucet toward the frozen place.

If you are not sure whether the frozen pipe has sprung a leak, take precautions to protect furniture or floors from water before you thaw the pipes. Move items that might be damaged from the area or cover them with plastic drop cloths. Have several pans and pails handy to catch the leaks as they appear. If you detect a crack in a pipe, put a rubber pad and clamp on it before the pipe thaws.

One way to thaw a pipe is "grandma's frozen pipe cure." You wrap the pipe in rags and pour boiling water over it. This method works but is slow and very messy. A heating pad wrapped around the pipe or a heat lamp or hair dryer aimed at it works much better. Except on plastic pipe, the quickest thawing is done with a propane torch equipped with a flame spreader nozzle (see the drawing). Never let the pipe get too hot to touch and be sure to protect the wall behind it with a piece of asbestos or sheet metal.

Overflowing Washers

If either your washing machine or dishwasher overflows, turn the dial to off. If the water keeps flowing, the electrical circuitry may be at fault. Stop the washer by pulling the plug, throwing the circuit breaker, or pulling the fuse in the house fuse panel. If it would be quicker in your case, turn off the water to the washer. The valve should be behind the washing machine or under the sink beside the dishwasher. If you can't find or reach the shutoff valve to the washer, turn off the main valve.

When the water stops flowing, check the drains to see if they are plugged with dirt, lint, or grease. If the drains are clean and the washer still overflows, it is not a plumbing problem but an appliance problem. Call an appliance service center.

Minimizing Flood Damage

When you detect a substantial leak or overflow situation, turn off the water quickly before trying to find or fix the problem. Dam doorways with rolled up blankets, rugs, or towels to isolate the water.

If the flood is upstairs, your first clue may be a dripping light fixture or a stain on the ceiling or wall. If a light fixture is involved, first turn off the water. Then turn off the electricity at the fuse panel. Drain the light fixture by removing any ball, dome, or cover and be sure it's dry inside before turning the electricity back on. Poke holes in the ceiling to allow all the water to drain out.

If a flood really gets out of hand, call the fire department—that's one of the things they're trained for.

FIXING LEAKS

Leaks in Pipes, Faucets, and Valves

You can temporarily repair leaking pipes in several ways. For a proper and lasting repair job, you will need to turn off the water and replace the damaged piece of pipe (see pages 30–31) or take the joint apart, clean it, apply new pipe joint compound, and put it back together. Here are some emergency measures to tide you over until you can get to the permanent repair.

To stop a pinhole leak in a hose or pipe, push a toothpick or pencil lead into the hole and break the end off. Dry off the surface of the pipe and wrap it several turns with duct tape or electrical tape.

A large leak can be stopped temporarily by covering it with a pad of rubber, like an old inner tube or kitchen glove. Compress the rubber over the leak with a piece of wood and C-clamps or hose clamps. Rubber pads and clamps manufactured and sold in sets for this purpose are available in some stores.

If a whole section of pipe is leaking because it is broken or rusted through, turn off the water and cut out the section. Replace it temporarily with a length of garden hose held in place with hose clamps.

If threaded pipe is leaking at a joint, you may be able to stop it by turning the pipe with a pipe wrench to tighten the joint. Or you can turn off the water to relieve the pressure, clean the joint with a wire brush, dry it, and apply two-part epoxy cement. The instructions will tell you how long to wait before you turn on the water again.

Leaky faucets and valves can often be fixed just by tightening the packing nut. If that doesn't do it, turn off the water and replace the washer or packing (see page 52). Be sure to close the sink drain when working on the faucets so dropped screws and other small parts don't end up in the P-trap.

Water Heater Tank Leaks

Any repair you can make to a water heater tank must be considered very temporary—in fact, it's better not even to try. If a leak develops, it means the tank is corroded and will shortly have many more leaks. The best solution is to get a new water heater promptly. See page 76 for how to select and install one.

Attempt a repair only when there is a critical need for hot water and it's not possible to get it anywhere else. If for some reason you must try to repair the leak, the first step is to find it. This will entail cutting through the outer metal shell with airplane snips and removing the insulation until you find the hole, which will be a tedious and difficult job.

If you find the hole, turn off the gas or electricity and the water to the heater before you start your repair and don't turn the heater on again until you are sure your repair is successful. Make a stopper for it with a rubber washer and a toggle bolt as shown in the drawing. Insert the toggle bolt through the hole—enlarge the hole if necessary to get it through—and then cinch down the bolt to squeeze the washer over the hole.

If the leak has put out the flame on a gas water heater or you find that the leak is inside the flue, you will have to forget it. There is no way to reach inside the flue to even attempt a repair.

Stopping a Small Leak

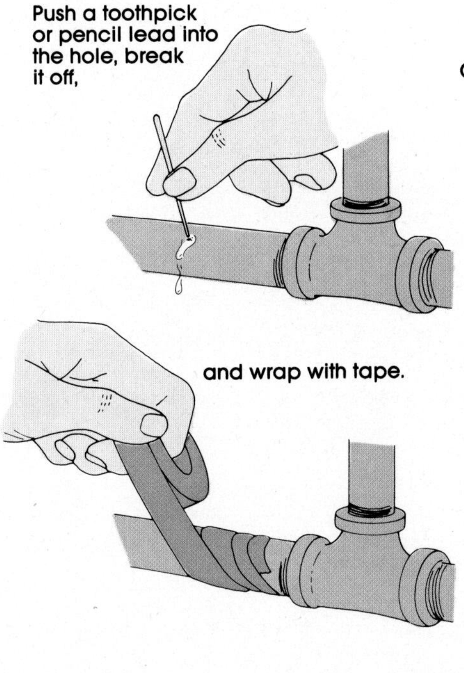

Stopping a Large Leak

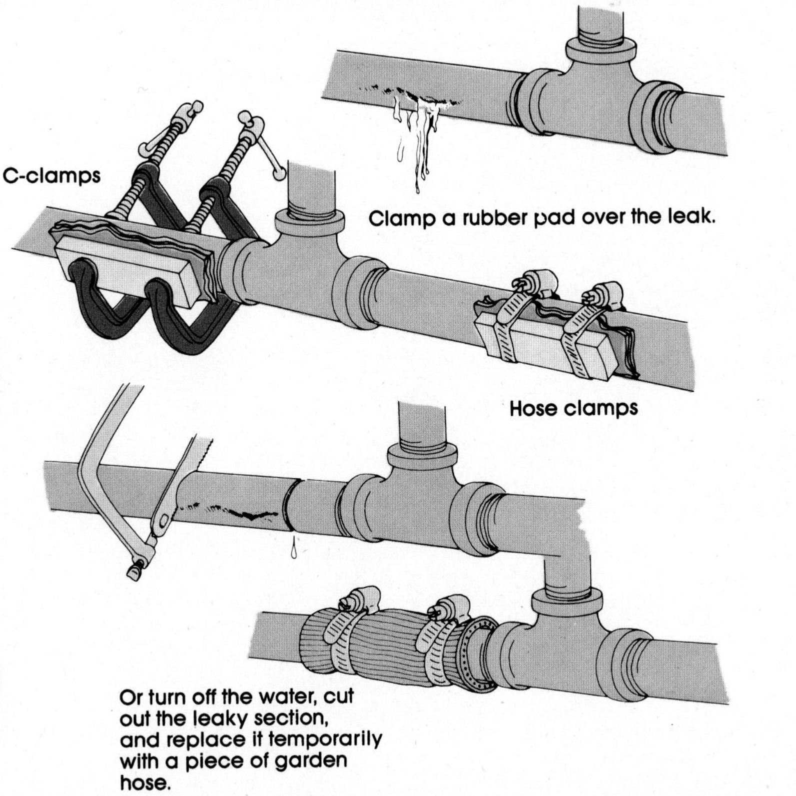

COLBY 3" ASTM F-628-79 COEX ABS

WORKING WITH PIPE

Should you use plastic pipe? Cast iron? Copper? What's involved in measuring, cutting, bending, joining, and repairing the various kinds of pipe, and what tools will you need for the job?

There are three basic ways to join pipe and fittings to each other. The first and probably the most common is with threaded or screw-together joints. The second is with fused joints, that is, with soldering, welding, brazing, or cementing. And the third is with compression joints, where the end of the pipe and the fitting are pressed together with a threaded nut, but the ends of the pipe aren't threaded. Flare joints in copper tubing and the connection of the tailpiece to a P-trap are examples.

Iron, brass, and old-fashioned copper pipe in standard or "schedule 40" thickness are almost always connected by threaded joints. Copper tubing and plastic pipe have some threaded fittings, but the pipe itself is almost never threaded. Copper tubing is usually soldered or held together with flare fittings. Plastic pipe is almost always welded to its fittings with solvent cement. PB (polybutylene) plastic tubing is joined with compression fittings; it cannot be solvent welded. Thinwalled brass pipe, like the 1¼-inch or larger chrome-plated drain pipe under your sink, is sometimes threaded but usually held together with flexible-ring compression joints. If it is threaded, the threads are very fine and were cut at the factory—you can't thread this kind of pipe yourself.

◀ As more kinds of pipe become available, you will need to know what can mix and what can't.

Some Words of Warning

Plumbing and electrical wiring don't mix. Your plumbing system, especially the metal cold-water pipes, is such a good conductor of electricity that it is used as a ground for your telephone and for all the wiring in your home. Be careful when working with your plumbing not to contact any electrical wiring, metal pull chains, or the like.

Ground connections. Look for an electrical ground connection on any pipes you are working on—it will usually be on a cold-water pipe (see drawing). If you must remove such a connection, replace it immediately on another cold-water pipe or in a different place on the same pipe. If it is a very large wire, it is probably the ground for your main supply panel. It is imperative that it remain attached at all times. Small wires may be from your telephone, washing machine, or another electrical appliance. It is not so urgent that these remain attached, but the appliance should not be used while the wires are disconnected.

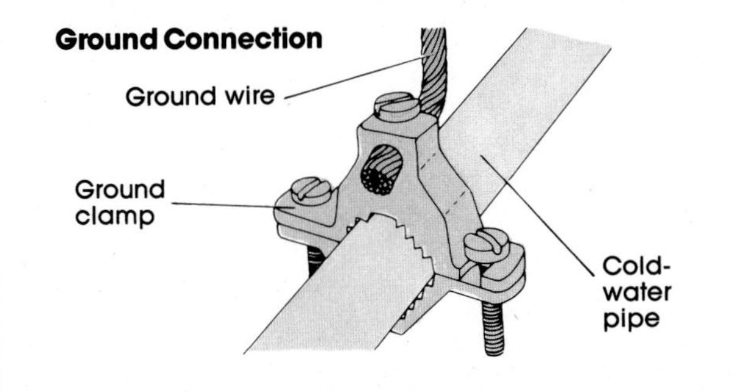

Galvanic action. When two different metals are immersed in water, a weak electric current flows between them. This is called a galvanic current. In plumbing installations, the two metals are usually iron—especially galvanized iron—and copper or brass. As the galvanic current flows from the copper, it carries copper atoms through the water and deposits them on the iron. Over a period of time, the copper pipe and fittings can deteriorate enough to cause leaks.

If you add copper pipe to an established system of iron pipe (or vice versa) you should put a *dielectric* fitting between the two unlike metals. These fittings have an insulator that prevents the iron from contacting the copper directly. Another method, preferred by some plumbers, is to install a brass fitting, a coupling or nipple, as an intermediate link between the iron and copper parts of the system. Although dielectric or brass fittings are very helpful, they cannot completely prevent a very slight, but unrelenting flow of electricity and copper atoms.

THREADED PIPE

Old-fashioned thickwalled copper and brass pipe have the same threads and fittings as galvanized steel pipe but are rarely used these days. Copper pipe is no longer available. Brass pipe is available but is so expensive that it is used only in small amounts for decorative purposes or to separate iron pipe from copper tubing to minimize galvanic action when the two metals are used in the same system.

Galvanized steel pipe, usually called iron pipe, has been used in more homes than any other kind. Almost all homes built prior to 1955, and many homes built since then, will have threaded iron pipe throughout. This kind of piping system has been used since before the turn of the century. Although the life span of galvanized iron pipe is usually considered to be 20 to 30 years, many homes built in the early 1900s have iron plumbing systems that are still being used. Most of these, of course, have had some parts of the system repaired or replaced, but many have a good deal of the original system still in service. Its ready availability, relatively low cost, and long life span are the main reasons for using iron pipe. Its strength and ability to withstand bumps and pressure make it useful in hostile environments, too.

It also has a couple of drawbacks. Iron rusts. Even though it is galvanized and the threads are covered with pipe compound when it is put together, eventually every system will need to have some sections replaced because they rust out. Another problem is that the relatively rough interior surface and the ridges where pipe and fittings meet—as compared to sweated copper or plastic pipe—cause mineral deposits from hard water to build up, which may eventually block the pipe altogether. The installation of a water softener will sometimes reverse the action, however, as the softened water tends to dissolve the deposits.

If you plan on doing some plumbing yourself, there are other negative attributes of threaded iron pipe. You will need a greater number of tools and a little more skill to work with it. Thickwalled iron is very unforgiving because of its rigidity. You will need to be much more accurate in your measuring and cutting than with the more flexible copper tubing or plastic pipe. If you cut a piece too long, you will have to recut and rethread it. If it's too short, you will have to replace it entirely.

Threaded pipe and fittings of all materials will have standard pipe thread. Whether pipe is threaded at a factory or hardware store or you do it yourself, it will all have the same threads and will always fit with other pipe and fittings of the same diameter. There are only a few places in a plumbing system where you will find any other threads. Thinwalled brass tubing used for sink drains will sometimes have a fine machine thread put on at the factory; compression fittings will have machine threads just like those on nuts and bolts (which is what they are), and faucets will often have the ends threaded to accept a garden hose. Hose thread is a very coarse thread found only on hose and fittings designed to connect to hoses.

Pipe thread is unique in that it is tapered. It is cut at an angle so the thread at the end of the pipe is smaller in diameter than the one a half inch up the pipe. The taper, which is about ¾ inch per foot, causes the joint to tighten as it is screwed together so it seals the joint.

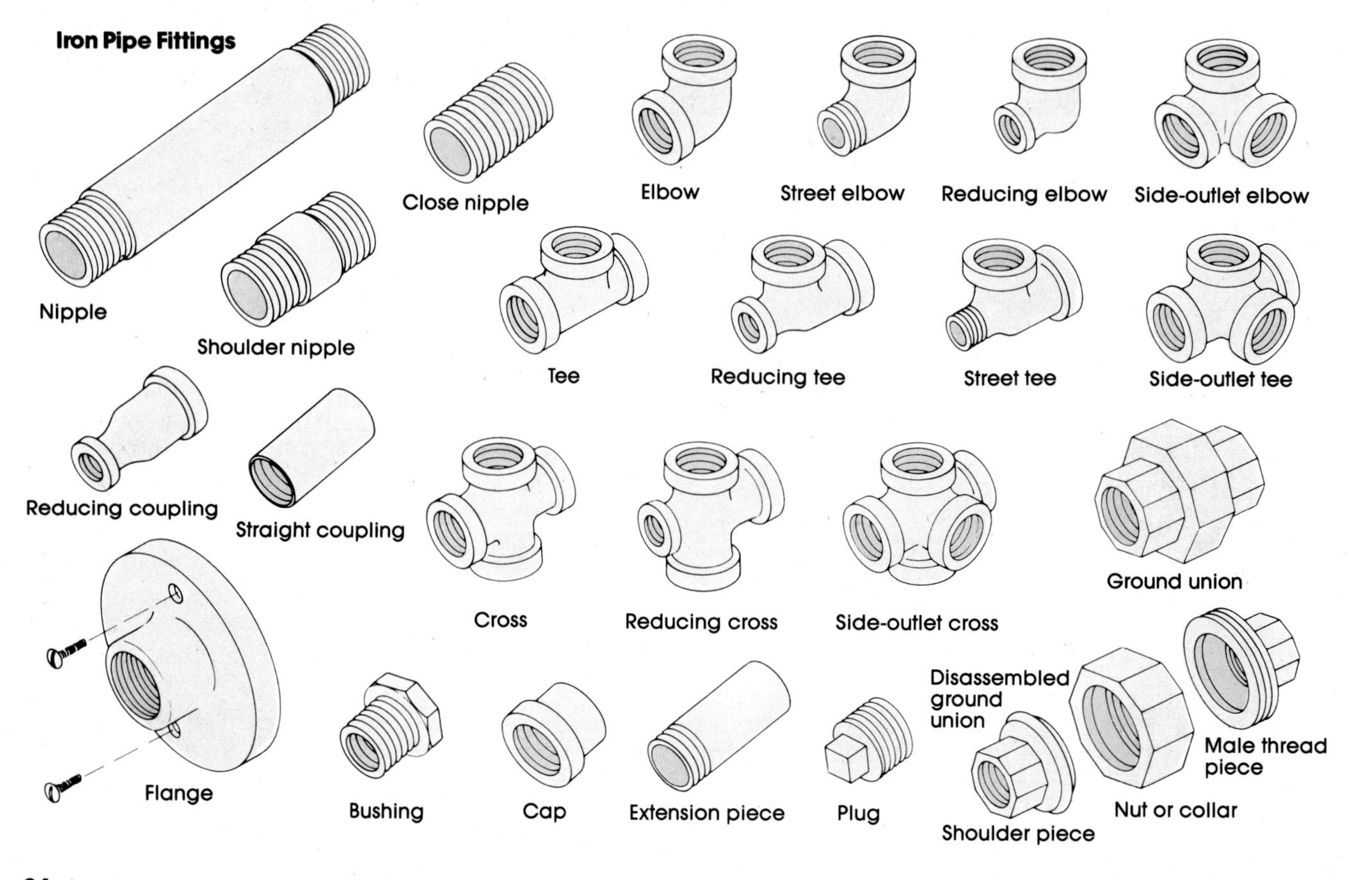

Pipe and Fittings

Iron pipe and fittings are manufactured in sizes from ¼ to 2½ inches. However, a hardware store or plumbing supply house that caters to homeowners rather than contractors probably will stock only the ones commonly used in houses: ½-, ¾-, and 1-inch and sometimes a limited quantity of 1¼- and 1½-inch pipe and fittings. The 2½-inch size is not usually stocked but can sometimes be special ordered. The pipe comes in standard lengths of 10 and 20 feet with threaded ends. Many dealers will cut lengths to order and therefore will have shorter lengths on hand if you want them.

Nipples. Originally *nipple* was a name given to any piece of pipe 6 inches or less in length. Now, however, plumbers and dealers often refer to any precut and threaded lengths of pipe they stock as nipples. Most stores have nipples in lengths of ½-inch increments from 1½ inches to 6 inches and in 1- or 2-inch increments up to 12 inches. Some stores even stock nipples of 18, 24, 30, and 36 inches. A nipple of 1½ inches or less that is threaded for its entire length is called a *close nipple.* One that has only a small unthreaded section in the middle—¼ to ½ inch or so—is called a *shoulder* or *short nipple.* These are obviously used when fittings need to be touching or very close to each other.

Elbows. Also referred to as *ells,* these are used when the pipe must change direction. They are available in 45-degree and 90-degree angles and in several forms. A standard ell has both ends the same size and both openings are female. A *reducing ell* has one opening smaller than the other. A *street ell* is female on one end and male on the other.

Tees. These fittings are shaped like a "T" and used when a pipe line branches at a right angle. A *straight tee* has three female openings, all the same size. The most common *reducing tee* has run-through openings the same size and the branch opening smaller, but they are manufactured in several combinations of two and three different-sized openings. The drawings show a few of them. A tee with two female openings and one male opening is called a *street* or *service tee.*

Crosses. When you want two branches to leave the main line at the same place, you use a cross. A *straight cross* has all the openings the same size. *Reducing crosses* come in several combinations of sizes, just like tees.

Side-outlet ells, tees, and crosses. All of these are just like regular ells, tees, and crosses except that they have one additional opening at a right angle to the plane of the others, as shown in the drawings.

Couplings. Couplings are used to join two pieces of pipe without changing the direction. In a *straight coupling,* both openings are the same size. In a *reducing* or *bell coupling,* one opening is smaller than the other. Usually couplings have female openings at both ends. One with a female opening on one end and a male opening at the other is called a *reducing piece.*

Unions. A union joins two lengths of pipe of the same diameter without changing direction. It is different from a coupling in that a union can be installed and removed without turning either length of pipe. A union is a three-part fitting. The center piece or nut is slipped onto one of the lengths of pipe. Then the two end pieces are screwed onto the two lengths of pipe, using pipe joint compound—the one without threads must go on the length of pipe that is holdling the nut. Line up the pipes, mate the ends of the union, and tighten the nut. The nut will draw the ends of the union tightly together. Use no pipe compound where the ends of the union join. The smooth mating surfaces of the union will seal tightly without it. Unions are generally used where a new branch line is added to an existing system or where an appliance or fixture must be replaced periodically. A water heater, for instance, is usually attached to both the hot- and cold-water pipes with a union.

How to Join Pipes with a Union

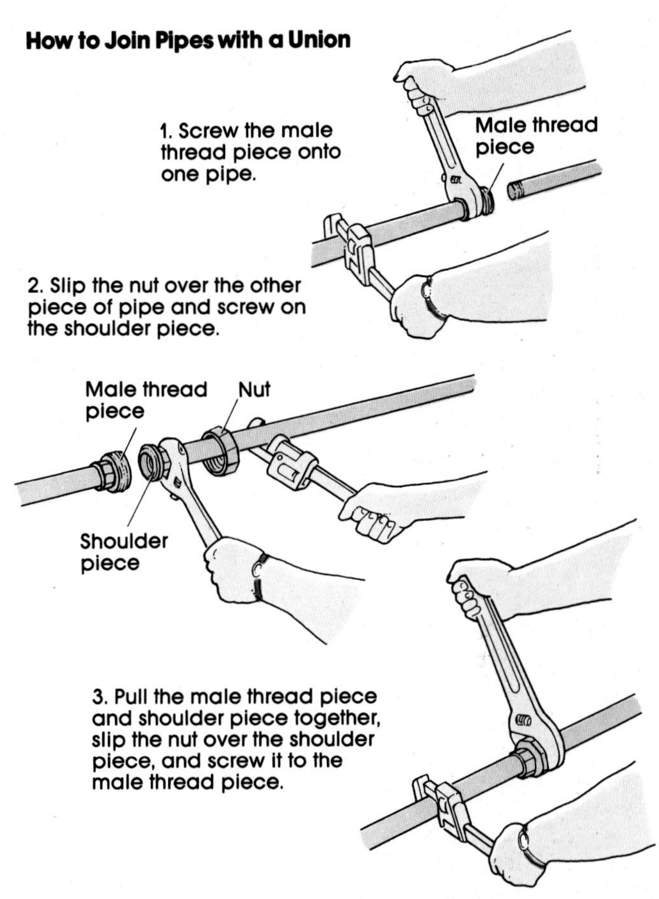

Bushings. A bushing is usually used to reduce one opening in a tee, elbow, or other fitting so it can accommodate a smaller pipe. It is both male and female; that is, it has threads both inside and outside and is hexagon-shaped on one end so it can be turned with a wrench into the fitting or onto the pipe.

Caps. A cap, or concave piece of galvanized iron with female threads inside, is used to seal the end of a pipe for testing or to make an air chamber. It can also be used to close off a pipe that is no longer in use.

Plugs. A plug is a solid piece of galvanized iron with male threads on the outside and a square head that can be gripped with a wrench. Some plugs have a slot like a screw, but these are rare. A plug is used to close one opening in a fitting when it's not needed. It can be used along with a coupling to seal the end of a pipe.

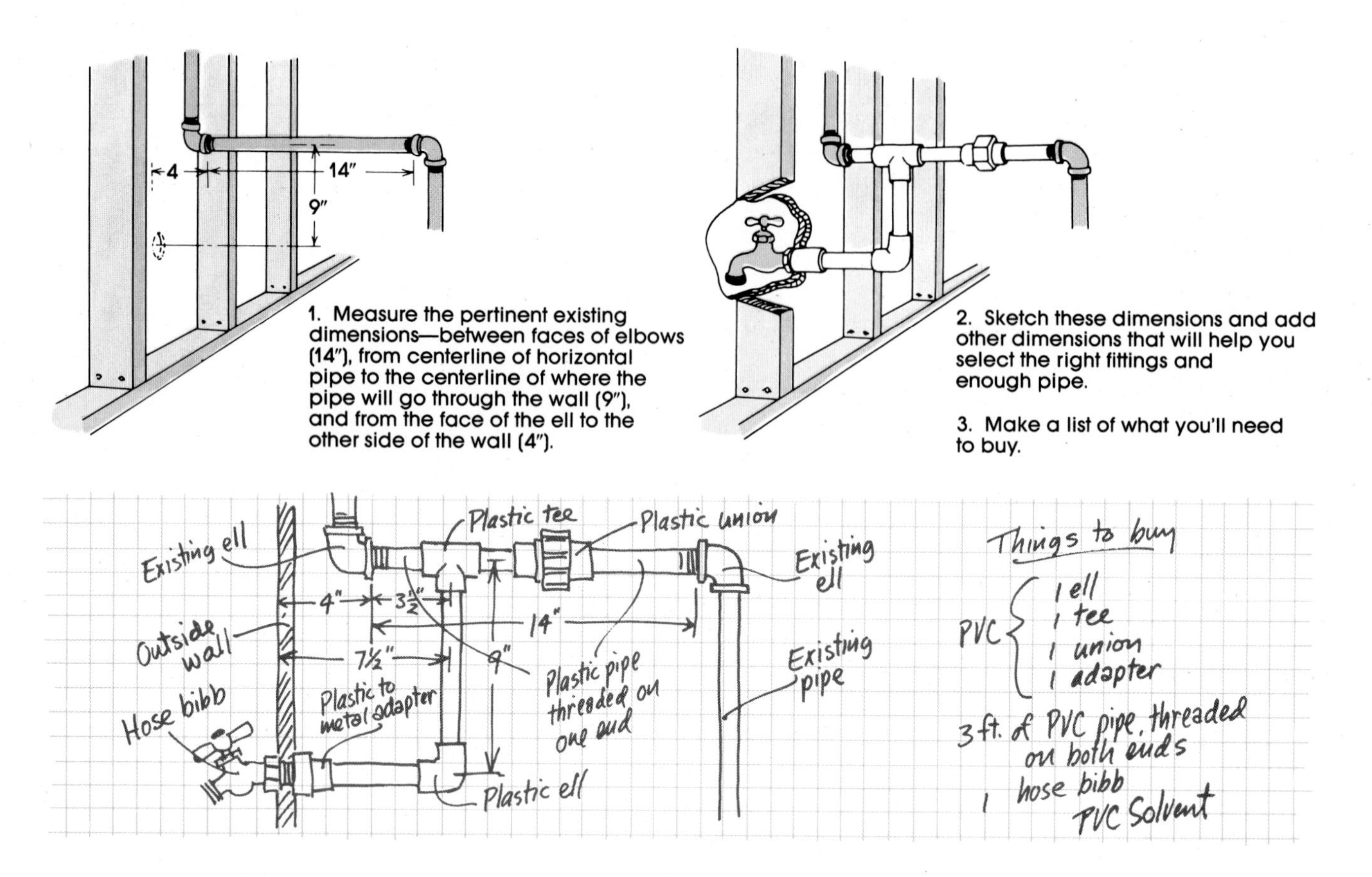

1. Measure the pertinent existing dimensions—between faces of elbows (14"), from centerline of horizontal pipe to the centerline of where the pipe will go through the wall (9"), and from the face of the ell to the other side of the wall (4").

2. Sketch these dimensions and add other dimensions that will help you select the right fittings and enough pipe.

3. Make a list of what you'll need to buy.

Measuring, Cutting, and Threading

Because iron pipe is so inflexible, precise measuring is essential to a neat and efficient job. If you plan to let someone at the hardware store cut your pipe for you, it's even more critical in order to save running back and forth to the store and to avoid buying a lot of extra pipe.

First sketch out the installation you are planning. It can be quite rough, just clear enough for you to follow. Mark on the sketch the fittings you'll need. Will you need a union? A tee? How many ells? In order to get accurate measurements, it would be a good idea to go buy at least one of each of the fittings you'll need and to use them to precisely determine their dimensions and that of the pipe.

For instance, if the new pipe must go from a mainline pipe to a wall and then turn and follow the wall, you will need to know how much distance will be taken up by the tee you'll put in the main line and the elbow at the wall. When measuring the length of the pipe, hold the fitting in place and mark the position of the face of the fitting. Measure the distance between the faces of the fittings and then add the length of the threads that will be inside each fitting. The thread inside the fitting will be almost exactly ½ inch if you are using ½- or ¾-inch pipe; 9/16 inch for 1-inch pipe; and ⅝ inch for 1¼-inch pipe.

You can cut and thread iron pipe at home very easily. You will have to buy a pipe cutter, a reamer, and a set of dies with a holder, also called a *stock*, with pipe handles to hold and turn the dies. A pipe vise is also a good idea if you are going to do very much cutting. If you are only going to cut and thread a few pieces of pipe, you can usually rent the cutter, reamer, threader, and vise. You can also modify a machinist's vise to hold the pipe (see drawing). If you only have one or two pieces

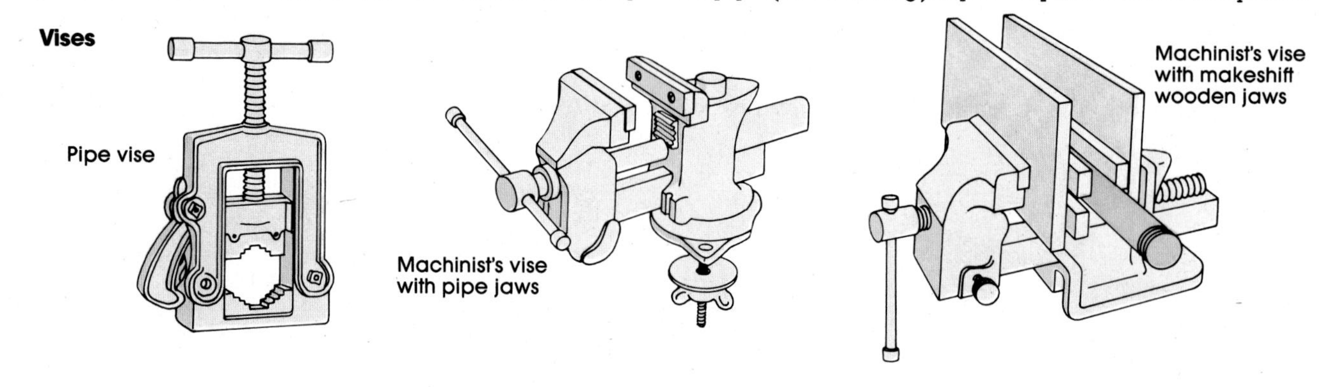

of pipe to cut and thread, let your local hardware store or pipe supply shop do it for you. As long as your measurements are accurate, this will cost the least in both time and money.

The dies for threading pipe are made of hardened steel teeth that cut the threads into the relatively soft steel of the pipe. Most die sets have cutters for pipe of ⅜-, ½-, ¾-, and 1-inch diameters, which are the sizes you'll need to work on any water or gas supply pipes in your house. If you need 1¼-, 1½-inch, or 2-inch pipe cut and threaded for a drain system, it's probably best to have it done at the hardware or plumbing supply store.

Make sure the piece of pipe you are going to cut is long enough. Check both ends to be sure the threads are not damaged. If one end is damaged, make your measurement from the undamaged end or recut the threads before you start measuring. Mark the pipe with a sharpened crayon or pencil where you want to cut. Yellow or white shows up best on the gray pipe and makes it less likely you will make an error.

To cut the pipe, put it in your vise with your mark about 6 to 8 inches from the vise. Open the cutter wide enough to clear the diameter of the pipe. Set it on the pipe with the cutting wheel on the mark. Apply cutting oil to the cutting wheel and rollers of your pipe cutter to make the cutting easier and to prolong the life of the wheel. Turn the handle so the wheel bites into the metal just a little bit. Rotate the cutter a full turn around the pipe to make a very shallow cut. Turn the handle a little more and rotate the cutter. Then do it again. You will soon get the knack of how deep to set the cutting wheel each time. Turning the handle too much can break the cutting wheel or spring the frame of the cutter. Once the pipe is cut off, wipe the end with a rag. Be very careful not to cut yourself on the sharp edge or on any burrs (rough spots). Check the cut for smoothness. Remove the burr around the inside of the pipe with a reamer and smooth any roughness with a file.

You can cut pipe with a hacksaw, but any jagged edge or deviation from a right-angle cut will make threading difficult or impossible. If you must use a hacksaw, make the cut just as square and smooth as possible and use a file to clean and smooth the end when the cut is completed.

To thread the pipe, insert the proper die in the diestock. Be sure the pipe is tight in the vise. Slide the diestock over the end of the pipe and apply pressure with the heel of one hand while you turn the stock slowly. When the die has taken a small bite so it is firmly started, put a lot of cutting oil on the die. Give the stock a complete clockwise turn; then turn it counterclockwise about a quarter turn. Again turn clockwise a full turn and back a quarter turn. This backing off a quarter turn for every full turn you make clears cut metal from the die and burrs from the new threads. Continue cutting in this manner until ⅛ to ¼ inch of the end of the pipe emerges from the diestock. Be sure there is always plenty of cutting oil on the die. When the cutting is finished, back the die off carefully to avoid damaging the new threads. Use a heavy rag to wipe away excess oil and chips.

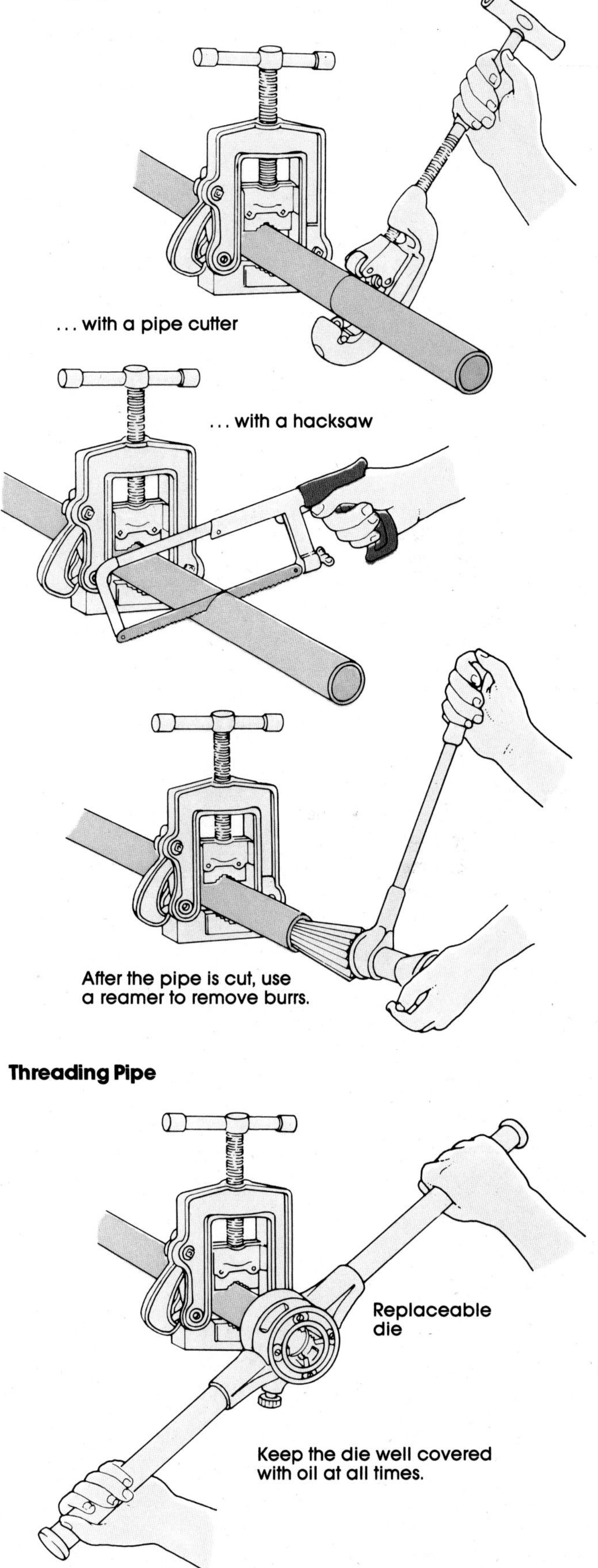

Joining Pipe and Fittings

Most professional plumbers assemble sections of the system they are installing at a workbench when they can. They think it is easier than doing it while lying on their back in a damp, cramped crawl space. You can usually screw at least one fitting to the pipe you just threaded while it is still in the vise. Then screw the assembled pipe and fitting into the last fitting you've already installed.

To get a good, leakproof joint, inspect both the male and female threads. Both the pipe and the fitting must be free of rust, burrs, and metal chips. Put joint compound or teflon tape on the male threads. Apply it evenly. Do not try to put compound or tape on the female threads. Screw the fitting onto the pipe by hand. It should go on easily for about three or four turns. Then turn it on snugly with pipe wrenches.

When assembling pipe, select wrenches of the proper size for the job. A wrench that's too small requires too much work, which causes unnecessary strain on your body and can cause slips that take skin from knuckles. A wrench that's too long may cause you to tighten the fitting too much. The result can be a cracked fitting or distorted threads and a leak. For putting together ½- to 1-inch pipe, a 12- to 14-inch pipe wrench is the right size; 1¼- to 1½-inch pipe requires an 18-inch wrench. Of course, taking apart an old system, if you are adding on, can require the biggest wrench you can find—old pipes are very stubborn.

To minimize strain on pipe and fittings as well as their supports, always use two wrenches. Place one on the pipe or fitting that is already installed to keep it from turning while you turn the other piece with the other wrench. A pipe wrench can be damaged if you use it the wrong way. You should always apply the pressure to turn a wrench in the direction of the open jaws. Wrenches must be set on the pipe in one way for assembling and the opposite way for taking apart, as shown in the drawing.

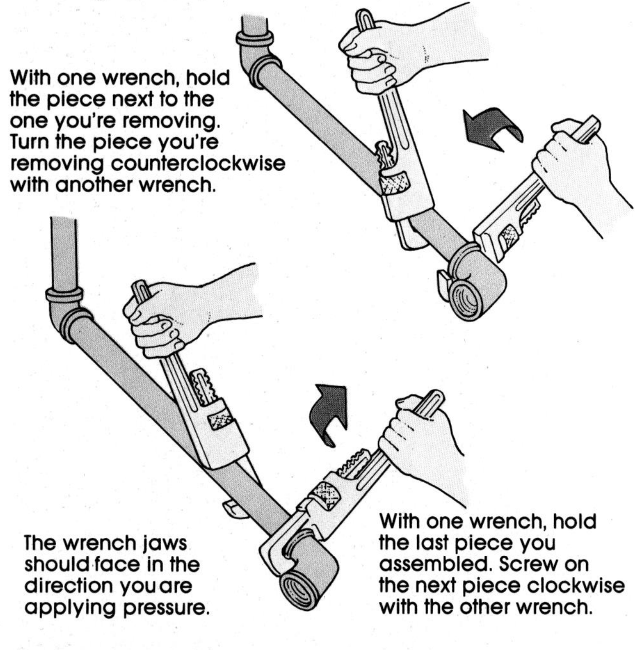

Repairing Iron Pipe

As described in the "Emergency Repairs" section on page 23, a small leak in a pipe can be temporarily fixed with a piece of rubber and a clamp. You can more permanently repair a small leak with a special pipe clamp available at plumbing supply shops.

Small leaks around joints can sometimes be repaired by turning off the water and packing the leaky place with epoxy cement. Let the epoxy dry completely before turning the water back on. If the leak persists, you will have to unscrew the pipe and fitting, clean the threads with a wire brush, apply a thick coat of new pipe joint compound, and reassemble the joint.

If the leak in the pipe indicates that it is rusted out and must be replaced or if the joint leak is in a place where it is impossible to unscrew one or two fittings, you will have to cut the pipe. This is really not as bad as it may seem at first—thanks to unions.

First turn off the water to this section of the system. With a hacksaw or pipe cutter, cut the leaking section of pipe or a section of pipe adjacent to the leaky fitting. The piece you cut will have to be at least 6 inches long in order to incorporate the union when it is replaced. Next unscrew the piece of cut pipe and, if you are repairing a leaky joint, clean and apply new compound as described. Reassemble the joint.

Next measure the length of pipe you need to replace, subtract the distance that will be taken up by the union, and divide the remainder into two parts, not necessarily equal (see the drawing). Screw the two sections of pipe into the fittings on each end and then install the union as described on page 27.

Tools for Working with Threaded Pipe

Pipe vise. This vise is made especially to hold pipe and other round objects. It is quite expensive and, unlike a machinist's vise or wood vise, cannot be used for a lot of other household jobs. Unless you expect to do a lot of plumbing jobs over the years, you may want to rent a pipe vise instead of buying one.

Pipe cutter, reamer, and threader. These are all the tools needed, along with the pipe vise, to cut and thread pipe yourself. These, too, are quite expensive to buy just for one little job. They are available for rental in most areas.

Pipe joint compound or teflon tape. This joint compound comes in cans, tubes, and sticks. The sticks are like big crayons that can be rubbed into the threads. Teflon tape comes in rolls like adhesive tape. The purpose of these items is to lubricate the pipe threads during the assembly of the pipe and fittings, to seal the threads so leaks are less likely, and to protect the threads from rust and other corrosion over the years.

Choose the material that is easiest for you to use. Joint compound has been around for a long time and has proven to be very effective. Teflon tape, the newest method of covering the threads, is quite easy to use and will protect the threads and prevent their rusting together for longer than the joint compound. We prefer the teflon tape for most applications.

Repairing Iron Pipe

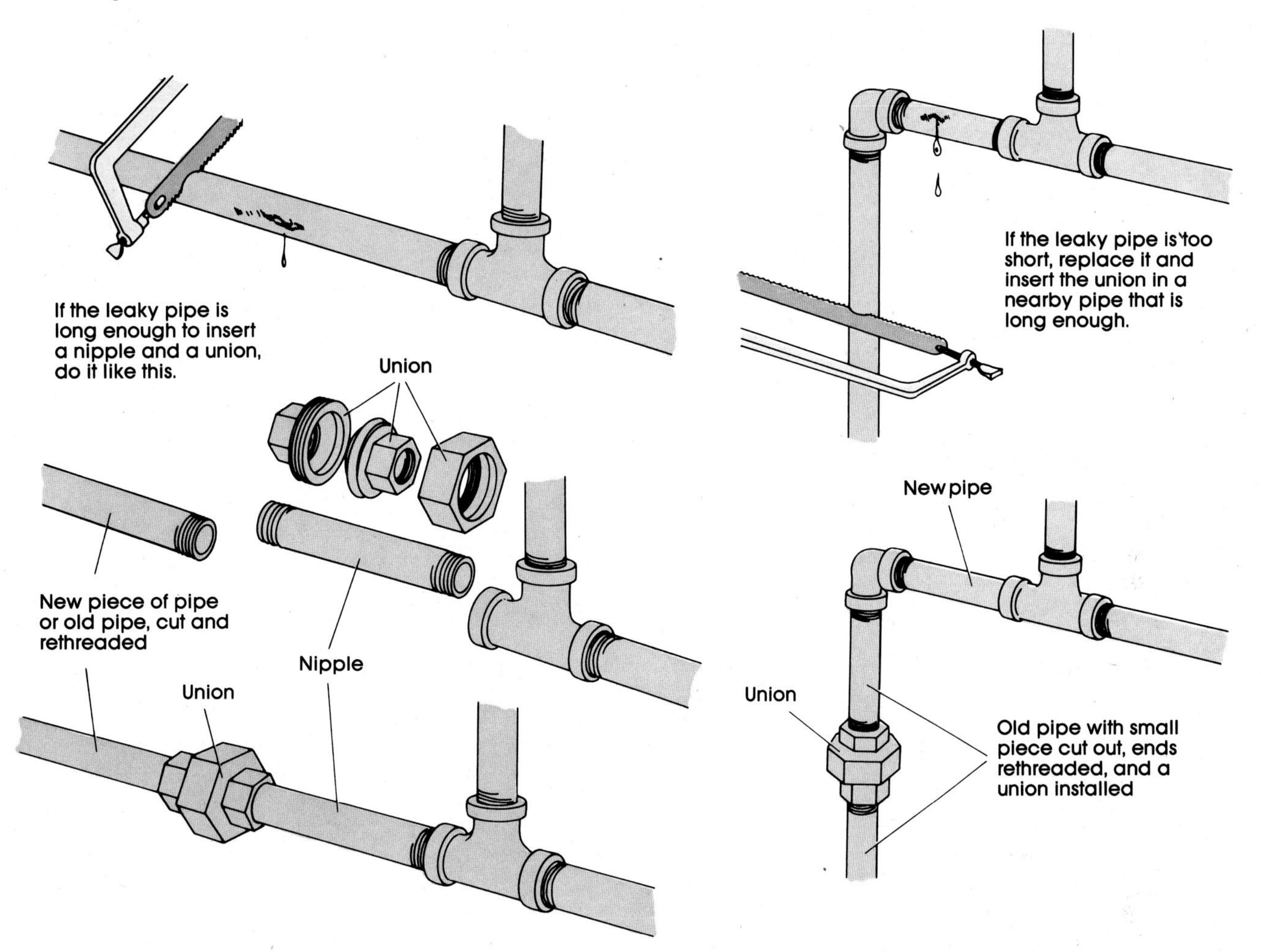

Threaded Pipe Tools

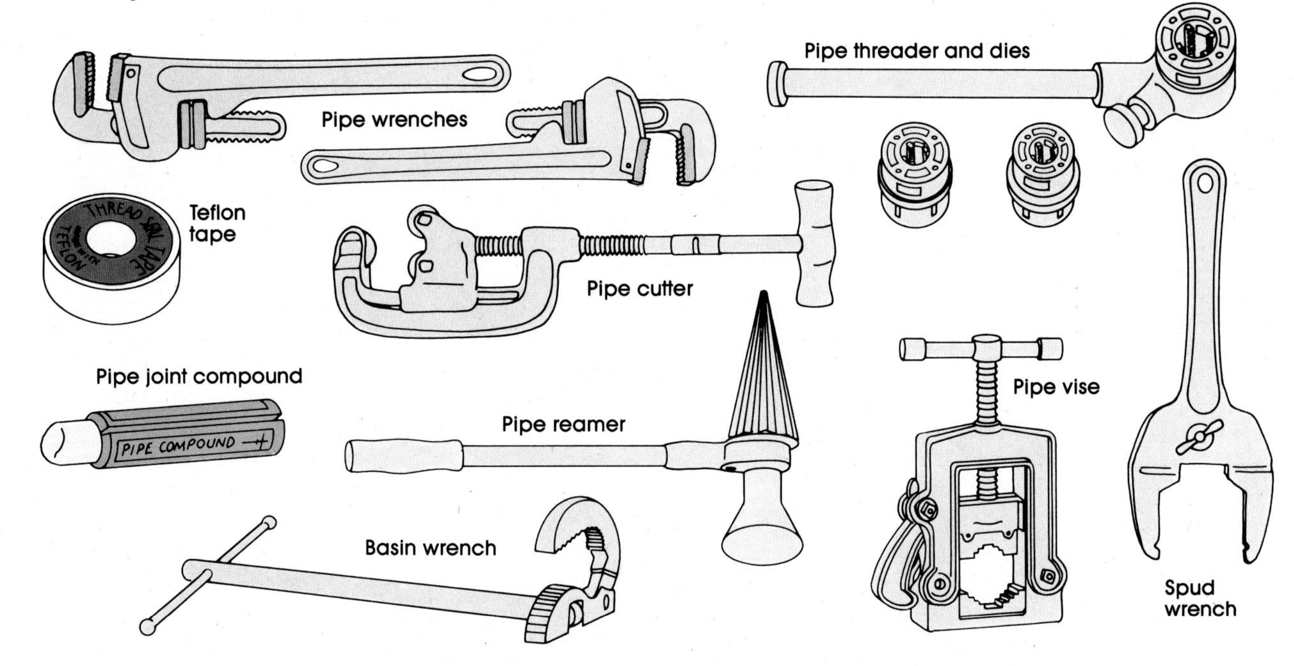

COPPER TUBING

The cost of copper tubing varies in its relation to the cost of iron pipe. Even when it costs more, however, the finished cost is usually less because it is so easy to work with, the labor cost or time for installing copper is much less than for installing iron pipe. Copper is lighter and therefore easier to carry and put in place than iron. For the same inside diameter, it is quite a bit smaller on the outside, allowing smaller holes in studs and floors and less clearance for joints and fittings. Because flexible tubing can be bent around corners, it takes fewer fittings than iron pipe and the fittings it does need are quicker and easier to install. The tools needed to install a copper plumbing system cost far less than those needed for a similar iron pipe system. Copper is also more resistant to corrosion than iron pipe in most cases, so you have much less trouble with it rusting out. Don't forget, however, that if you are thinking of installing a copper addition to an iron pipe system, you must consider galvanic action (see page 25).

Pipe and Fittings

Two kinds of copper tubing are available. Soft tubing, which is easily bent, comes in straight lengths of 20 feet and coils of up to 100 feet. Hard tubing is rigid and must turn corners using elbow fittings just like iron pipe. It comes in 10- and 20-foot straight lengths.

Copper tubing comes in the same sizes as iron pipe—from 1/4 to 2 1/2 inches. The actual outside diameter (O.D.) of the tubing is 1/8 inch larger than the nominal size, and the inside diameter (I.D.) varies with the thickness of the tube wall. Copper tubing for supply lines comes in three wall thicknesses, designated by the letter K for the thickest, L for medium, and M for the thinnest. Unless specified otherwise in your local building or plumbing code, M is usually considered adequate for home water supply systems. Copper drainage tubing comes in DWV thickness as rigid tubing only and only in the larger sizes used in DWV systems.

Another attribute of copper that can save you money is its superior water-carrying capacity. Because of the interior smoothness of both the tubing and the fitting connections, you can often use one size smaller pipe than with iron pipe and retain the same water flow.

The fittings for copper tubing are similar in most ways to those of iron pipe (see page 26). There are ells, tees, crosses, couplings, and reducers. In addition, there are adapters to connect the end of a threaded pipe to a copper tube.

Copper Fittings

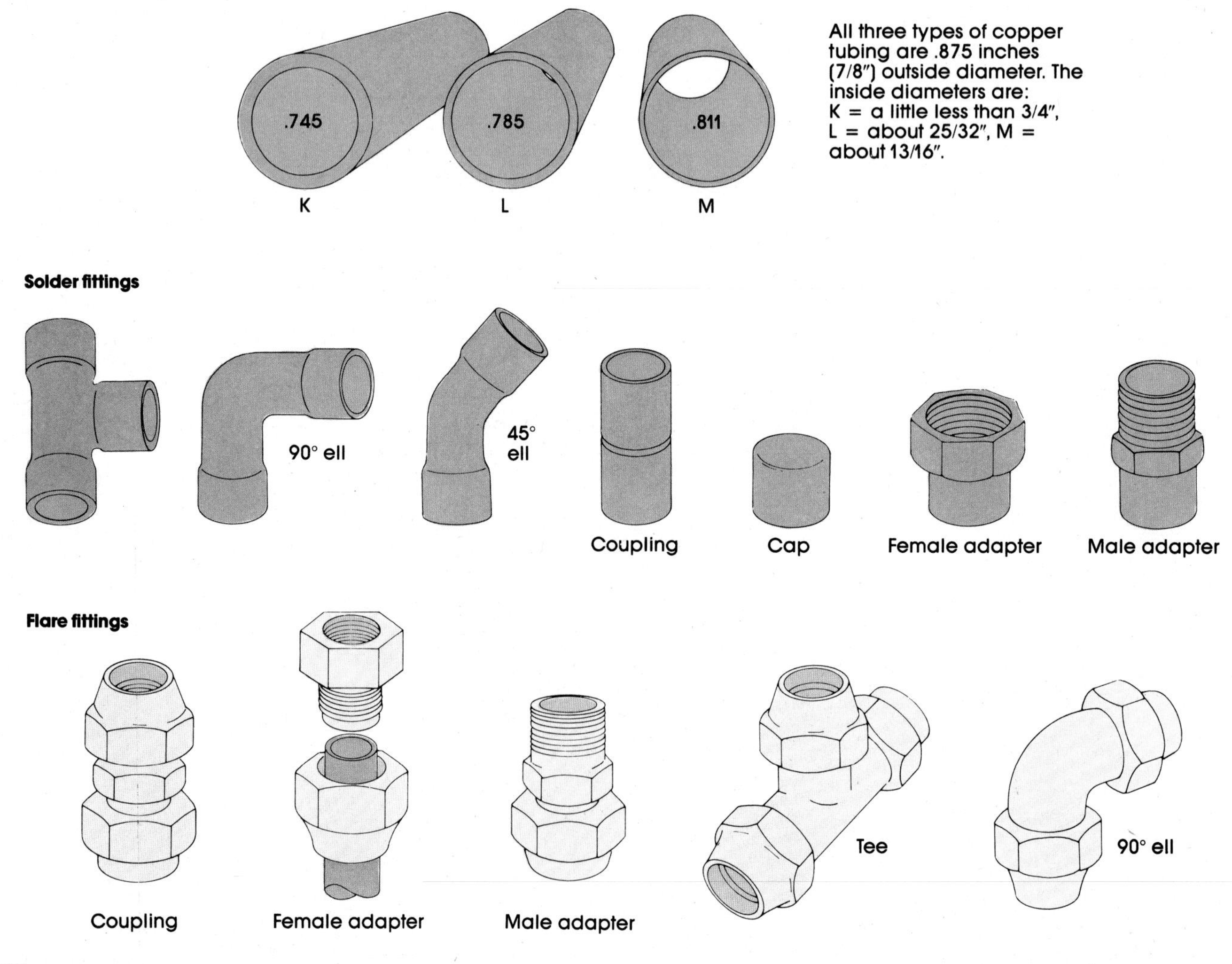

Bending and Cutting Copper Tubing

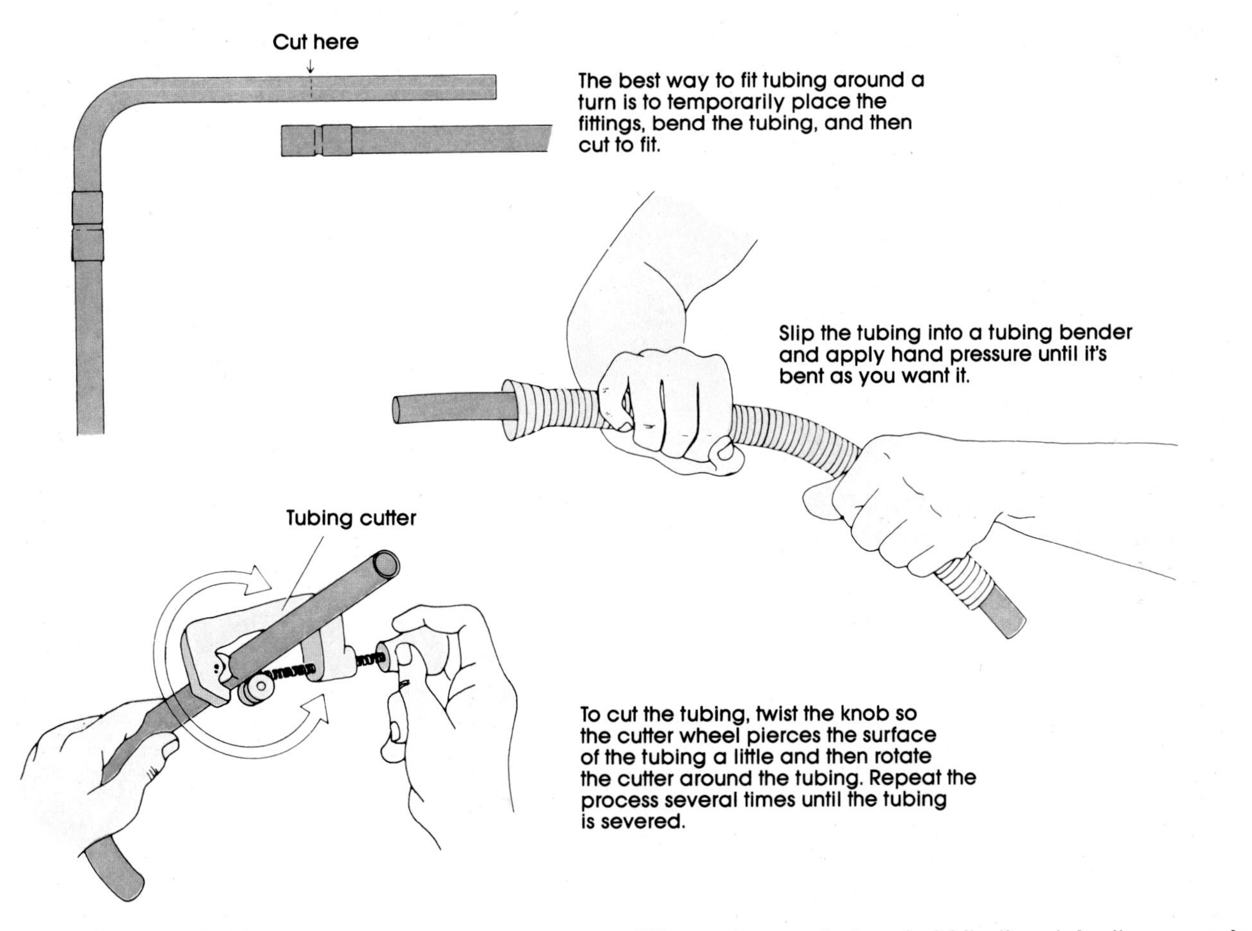

Bending Copper Tubing

Flexible or soft tubing, especially in the smaller diameters, can be bent without a tool if you are very careful and don't try to bend it too sharply. To be sure that your bends are without kinks or flattened spots, it is much better to use a tubing bender. A thinwalled tubing bender is a tube made of a tightly wound coil of hardened spring steel. It is usually about a foot long and comes in a range of diameters to accommodate the most common sizes of tubing. Just slip the tubing into the bender of the proper size and apply pressure by hand until the bend is as you want it. There are other styles of benders with grooved wheels and handles, but the wire-spring tubing bender is the simplest and is accurate enough for the smaller sizes of tubing.

Measuring and Cutting

Measuring a length of copper tubing is in some ways easier than measuring a length of iron pipe (see page 28). You still must measure the length between the faces of the fittings and add the amount of pipe that will go into the fitting. With copper tubing, however, you can actually slip the fitting onto the pipe to help you measure accurately.

If you are working with flexible tubing, it is a good idea to route the tubing where you want it before cutting. Just secure the starting end temporarily—insert it in a fitting or have a helper hold it—then take it over, under, and around to where you want the other end. You can usually even thread it through holes if necessary without cutting, especially if you have someone help. When you reach the place you want the end to be, hold a fitting against the tubing, mark it with a pencil or crayon, and cut it. It is usually a good idea to cut the tubing a little long for added flexibility in the final placement of the tubing and its supports.

Tubing can be cut with a hacksaw, but a tubing cutter is much quicker, easier, more accurate, and not very expensive. A tubing cutter works just like the pipe cutter described on page 29. Turn the handle to adjust the cutting wheel so it just rests on your mark. Turn the handle enough to press the cutting wheel into the tubing and rotate the cutter around the tubing. Repeat the process several times until the tube is cut off. The first cut or two will give you a feel for how much to turn the handle each time. Don't apply too much pressure or you will distort the tubing. When the tubing is cut, remove the burr around the inside with a reamer. Almost all tubing cutters come with a retractable or fold-away reamer built in.

If you must use a hacksaw, get a fine-tooth blade, 32 teeth per inch, and a small miter box. Be sure you make your cut smoothly and squarely across the tubing. Use a reamer or small round file to remove any burrs and to smooth the edge.

Joining Copper Tubing

There are several ways to join copper tubing. Sweating, tinning, and swaging are all used on rigid tubing and can be used on flexible tubing. Flare and compression joints are used only on soft flexible tubing.

Sweating. Sweating is a method of soldering that makes a very strong joint between copper tubing and fittings. If done right, the joint will be stronger than the tubing itself. Sweating is an easy process, but it does take a little practice to develop a feel for the proper temperatures.

Just a word of caution before you begin. If you are working on an existing plumbing system that has had water in it, open at least one nearby faucet and leave it open until you finish soldering. Even a few drops of water in the tubing you heat will turn to steam and expand greatly. If the part of the system you heat has no opening, the steam pressure could make one by bursting a tube and possibly injuring someone nearby.

If you are working within a few inches of a valve or faucet, take it apart and remove the washer and any other nonmetallic parts. The heat from your torch can melt, burn, or distort these parts if you don't. Also be sure that the fitting and the tubing have good support when you solder. Do not attempt to hold either in place with your hand. Even if you hold the tubing a foot or more from where you apply the heat, you may get burned.

To sweat copper tubing, you will need fine emery cloth or steel wool, a container of noncorrosive paste flux and a brush to apply it, lead-free solid solder wire, and a propane torch. Use the emery cloth or steel wool to clean and brighten the end of the tubing and the inside of the fitting. Apply a thin coat of flux to both brightened surfaces and slip the tubing into the fitting. Rotate them to distribute the flux evenly and wipe away any flux that has oozed out of the joint.

Evenly heat the fitting and the tubing near it with the propane torch, moving the torch back and forth and around. When the flux bubbles and smokes a little, remove the heat and quickly touch the end of the solder wire to the joint. If the fitting and tubing are the right temperature, the solder will melt and be sucked evenly into the joint by capillary action. The capillary action will even suck it upward if that is the way the fitting is positioned. If the joint isn't hot enough, the solder won't be sucked into the joint. If you use too much heat, especially more heat after you've applied the solder, some of the solder will run out of the joint. In either case, the best thing to do is to heat the joint to melt the solder, pull it apart, and start all over. Once you have a smooth line of solder all around the joint, it is filled and finished. Let it cool and go on to the next joint. If you are in a hurry, you can wet the joint with water to cool it, but don't touch or move anything until the solder loses its shine and you're sure it has hardened. This should only take ten seconds or so even if you don't wet it, but it will still be very hot to the touch.

■ CAUTION: When working with a torch in tight places, always have a piece of sheet metal or asbestos handy to protect the wood from the flame. It is also a good idea to have a plant sprayer or bottle of water to wet down any wood you may scorch. Always recheck and rewet any charred wood before you leave for lunch.

Sweating Copper Tubing

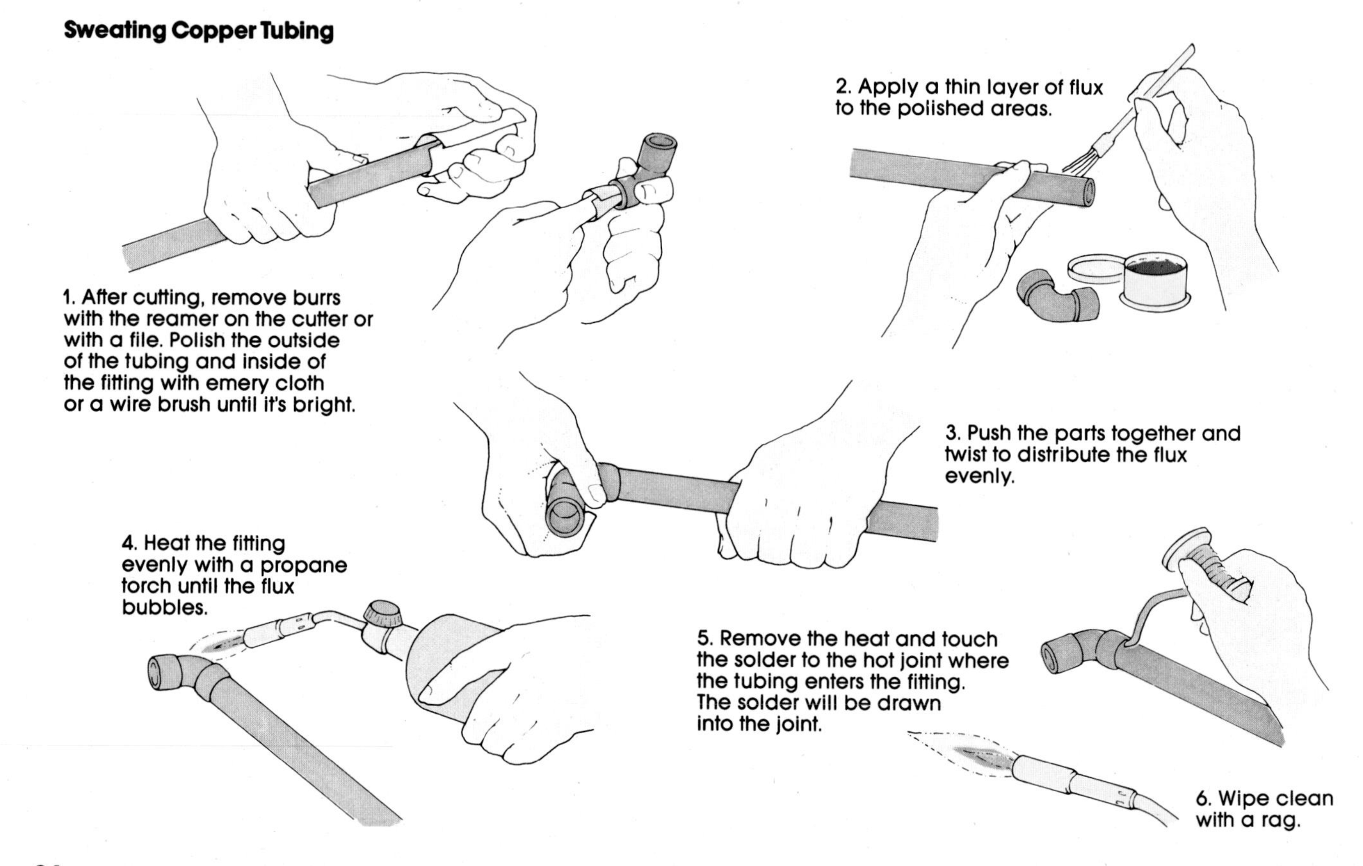

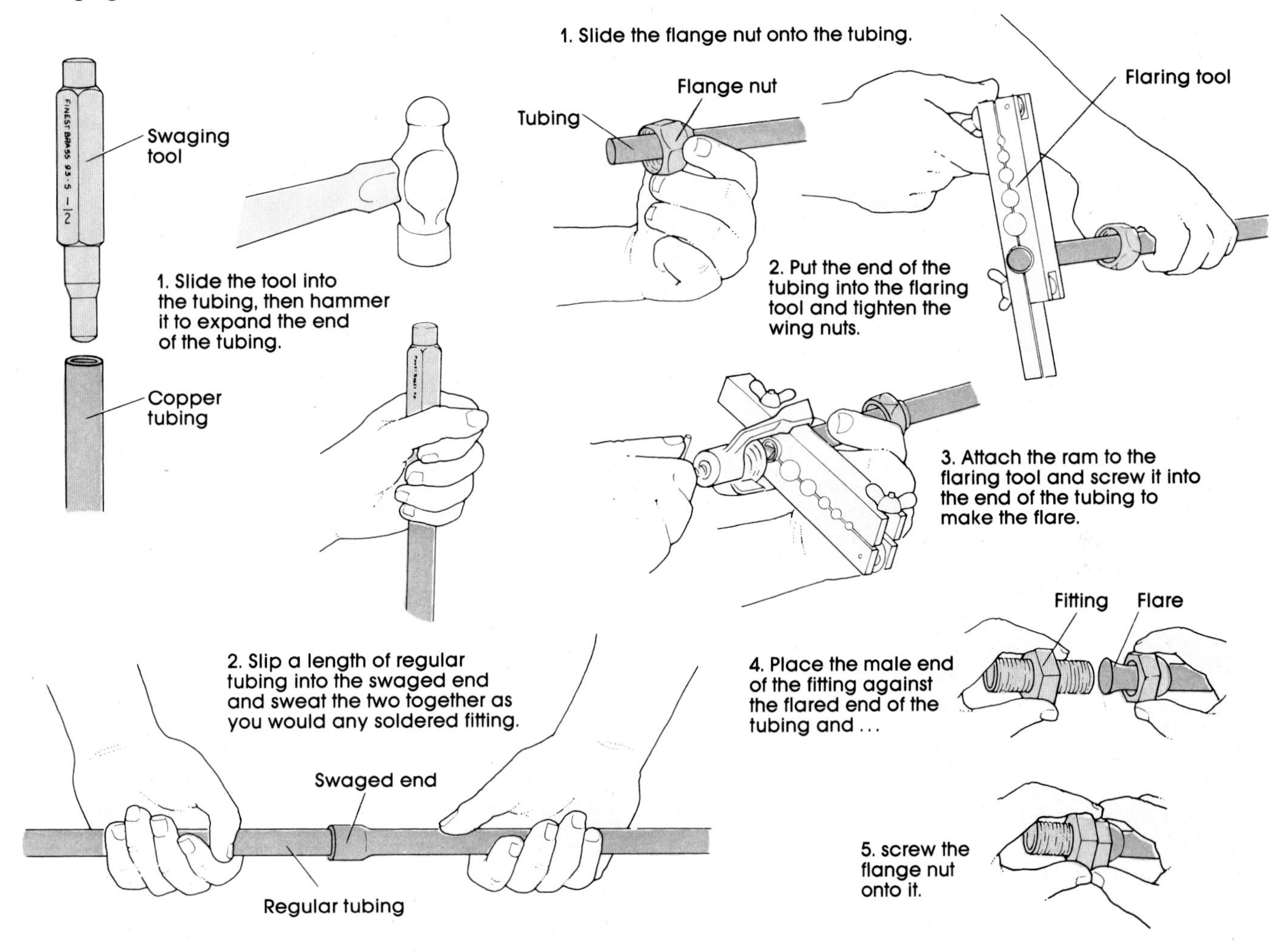

Tinning. This method is used when you are working on pipe above your head, like in a crawl space, and are afraid that dripping solder or hot flux will fall on you. Brighten the surfaces just as you would for sweating and apply the flux. Then heat each piece and apply the solder wire so you cover the contact areas thinly and evenly—cover both the outside end of the tubing and the inside of the fitting. Let each piece cool.

Then press the fitting into the tubing just enough to hold it. Apply pressure to the fitting or the tube, whichever is easier, with a board or pair of long-handled pliers as you apply heat. The instant the solder reaches melting temperature, the pieces will slide together with little chance for hot solder or flux to drip out. Remove the heat quickly as soon as you are aware that the solder is melting.

Swaging. This is a way of joining two pieces of thin-walled copper tubing without a fitting. In making a swaged joint, you expand one piece of copper tubing so another piece can slide into it. You will need a special swaging tool for this, and each size of tubing needs its own tool. Swaging tools usually come in sets for several common sizes of tubing. To form a swage joint, you place the swaging tool at the end of a piece of tubing and drive it in with a hammer or with a screw-down mechanism similar to a flaring tool. When you remove the tool, the tubing has an enlarged end just the size of a regular copper fitting (see the drawing). Now you are ready to sweat or tin the joint.

Flare joints. A flare joint can only be used on soft flexible copper tubing and only where the joint will not be concealed within a wall. Flare fittings are not usually recommended for extensive installations, but are useful in places where you might not want to use a propane torch. They are often used to connect fixtures or appliances that can be replaced easily at some future date.

To make a flare joint, you will need a flaring tool and the necessary fittings. Cut and ream the tubing as you would for a sweated joint. Slip the flange nut, the female part of the fitting, over the end of the tubing to be flared with its threaded end toward the end of the tubing. You won't be able to put it on the tubing after it is flared. Flare the end of the tubing with the flaring tool. Slide the flange nut against the flare, hold both against the male threads of the other part of the fitting, and screw the nut on as tightly as you can by hand. Then tighten it with two smooth-jawed wrenches and test it for leaks.

Compression joints. This joint is similar to a flare joint, but you need no special tools. It uses a rubber, plastic,

Compression Joining

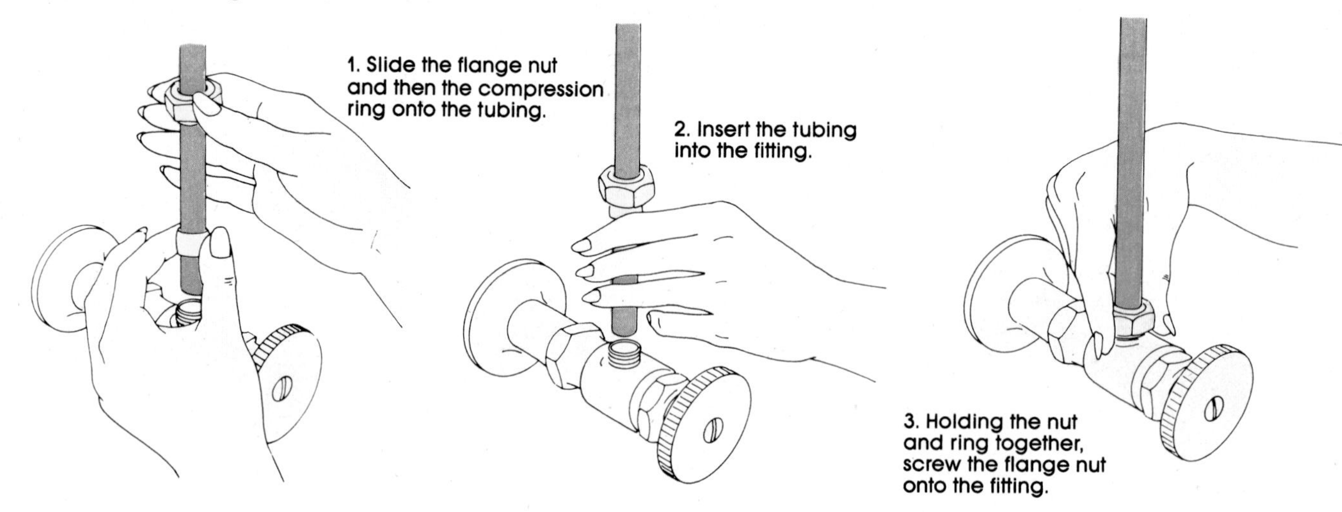

or metal compression ring instead of a flare in the end of the flexible tubing. Cut the tubing as usual, slide the flange nut and then the compression ring onto the tubing. Hold the flange nut and compression ring together at the end of the tubing and put the flange nut over the male end of the other part of the fitting. Tighten it with your fingers and then with two smooth-jawed wrenches.

You won't often need elbow fittings because the flexible tubing bends around corners. Other than that, flare and compression fittings come in all the usual configurations—tees, wyes, crosses, reducers, and so forth—as well as shutoff valves and other items.

Repairing Copper Tubing

A small pinhole leak in a copper tube or pipe is easy to repair with solder. Turn off the water and drain the line. Be sure to leave a faucet open in the line you are working on. Clean the area around the leak with emery cloth or steel wool. Apply a coat of flux. Heat the defective area with a propane torch and apply a dab of solder over the hole. This is only satisfactory for a very small hole—water pressure will pop the solder from a large hole.

For larger leaks, cut the tubing on each side of the hole. If the tubing can be bent or moved enough, clean and flux the ends and join them with a sweated-on coupling. If the ends cannot be pulled together, cut a small piece of tubing to fit between the ends and use two couplings.

It is not possible to solder any joint with water in it, even the slightest trickle or drip. In the event that closing the valves doesn't shut the water off completely, stuff a piece of bread into the tubing and push it several inches above where you want to solder. This temporary dam will hold back the water long enough for you to solder the joint and then will dissolve and wash away completely when the water is turned on.

One or more compression unions can be used instead of sweating on couplings if soldering is not desirable (see the drawings).

Repairing a Large Leak in Copper Tubing

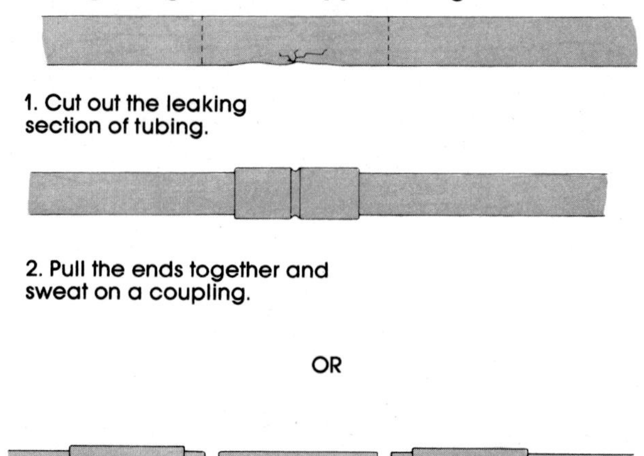

OR

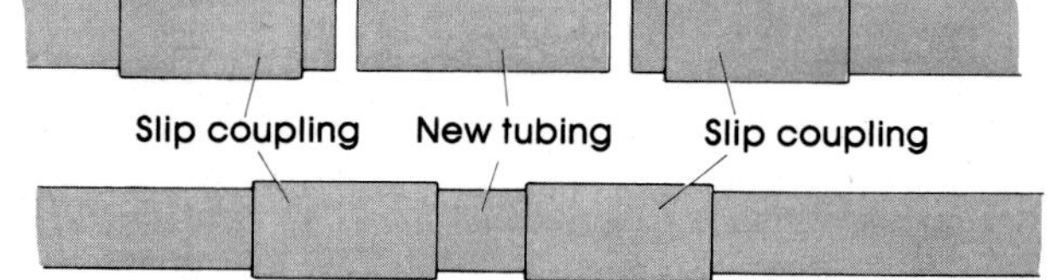

2. If the ends won't come together, insert a section of new tubing and sweat on two slip couplings.

OR

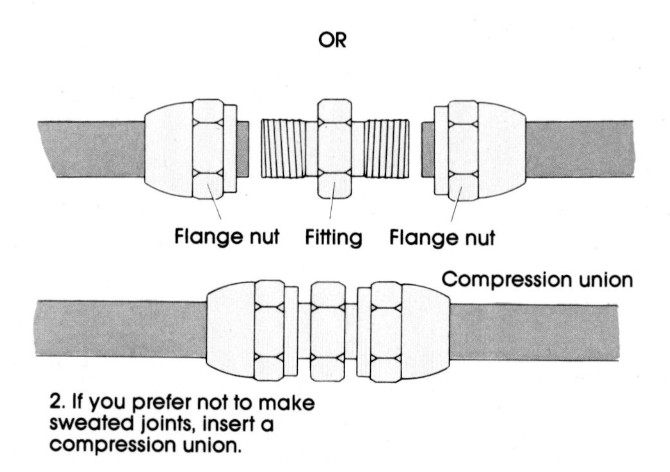

2. If you prefer not to make sweated joints, insert a compression union.

Tools for Working with Copper Tubing

Tubing bender. Here you can choose between two types. We prefer a set of the tightly wound spring steel benders. They usually come in a set of six for bending 1/4-, 5/16-, 3/8-, 7/16-, 1/2-, and 5/8-inch thinwalled tubing. You just insert the tubing into the bender and bend it however you want by hand. You can bend the tubing to a small or large radius just by how you manipulate it with your hands.

The other type of bender has a groove for each size tubing around a cylinder about 2 inches in diameter. You can bend the tubing as much or as little as you want, but the diameter of the bend will always be the same as the cylinder of the tool.

Tubing cutter. Using this tool is by far the quickest, easiest, and neatest way to cut tubing. Most types come with a reamer attached.

Steel wool, wire brush, or emery cloth. You'll need this to polish the surface of the copper for sweating. Because it conforms to the shape of the surface better, we think steel wool does a better job than emery cloth. It is less expensive, too.

Propane torch. The best kind to get is just a nozzle with a flame-adjustment knob that screws to a propane cylinder. This torch is great for thawing frozen metal pipes if it is fitted with a flame spreader. The spreaders are often sold with the torch.

Solder wire. Use solid solder wire that is labeled lead-free for joints in any system that supplies drinking water. For other applications use a solder that is acceptable to local building codes. The solid wire does not have a core of acid or rosin—for sweating copper it is better to apply paste flux yourself with a brush.

Noncorrosive flux and a brush. The main purpose of flux is to prevent the metal from oxidizing when heat is applied. This gives the solder a firm metal surface on which to adhere. It also cleans old corrosion from the metal if you missed some with the steel wool. The brush makes the job so much easier that it is well worth the few extra cents that it costs.

Sheet of asbestos or metal. This is to place behind the tubing you are sweating to prevent the torch flame from igniting the studs or joist behind the pipe. You only need this protection if you are soldering within 6 inches or so of the wood.

Swaging tool. This tool shapes the end of copper tubing so it will slide over the end of another tube of the same size (see page 35). It is not often used by the homeowner for small jobs.

Ball peen hammer. You'll need one of these if you opt to use a swaging tool. It has a different temper than a carpenter's hammer and is much less likely to chip when used to hammer on metal.

Flaring tool. You'll only need this if you prefer making copper tubing joints without solder. It flares the end of a copper tube so it can be held by special solderless fittings (see page 35). (This tool, along with standard flare fittings, can also be used to connect CPVC and PB plastic pipe to a metal pipe system).

Tools for Copper Tubing

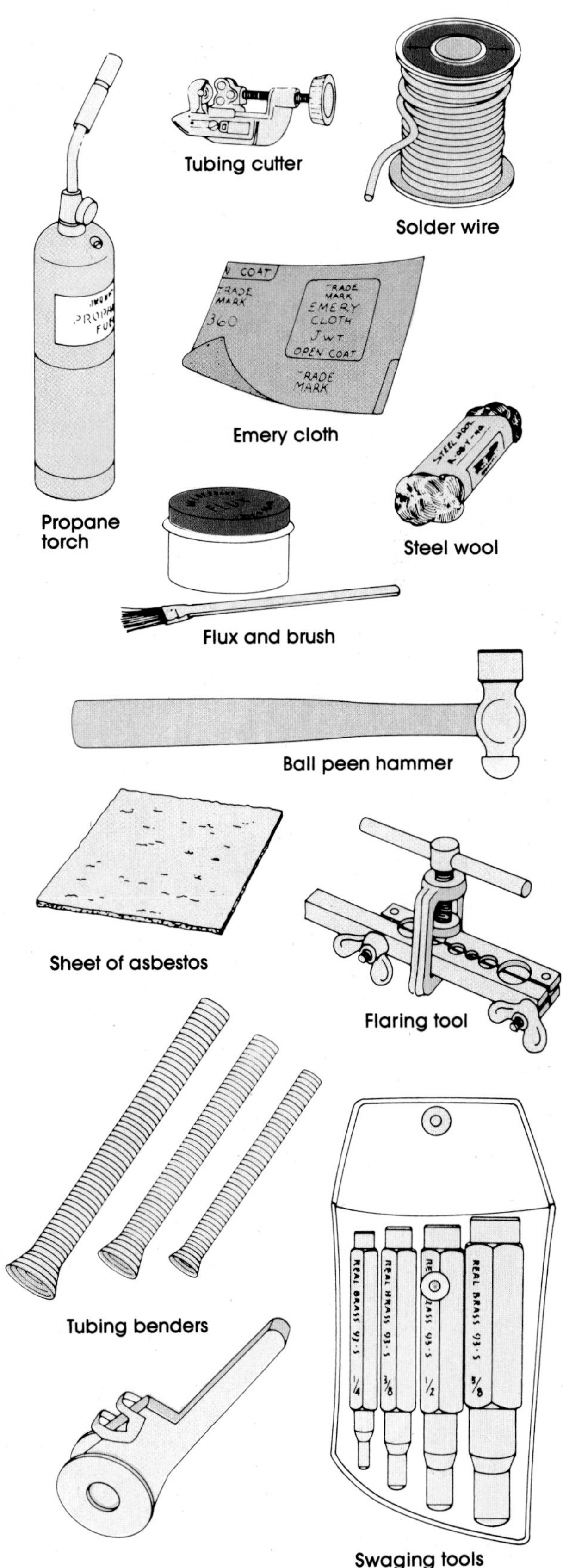

PLASTIC PIPE

If your local building or plumbing code permits, plastic is the material to choose for any new installation or addition to your plumbing system. It is lighter to move and carry about and much easier to join than any of the metal pipes. You use a solvent cement or compression fittings and the joints are just as permanent and watertight as any metal piping. You don't need expensive special tools or have to use potentially hazardous flame or molten metal.

Plastic can be cut with almost any saw or a tubing cutter. The flexible plastic pipe can be cut with a knife. Because plastic is chemically and electrically inert, it can be used in the same system with iron or copper without the threat of galvanic action. And it will not rust or corrode. Because it is so smooth inside and chemically inert, it will not collect mineral deposits as readily and clog up like metal. Another very good reason to use plastic is that it is less expensive than any metal.

On the negative side, plastic is prohibited for water supply systems in many localities. Some say that is because plastic would melt or burn in a fire, others say it hasn't been around long enough to be proven, and still others say labor unions oppose it on the grounds that it takes far fewer hours to install and would provide fewer jobs. In any case, if your local code prohibits plastic pipe, don't use it. The building inspector could make you remove it all and replace it with pipe that does meet the code.

Plastic pipe has other disadvantages. Plastic pipe or fittings cannot be used to support the weight of a fixture. Also, plastic expands considerably and tends to soften and sag when exposed to hot water for an extended period and not supported. Closely spaced, loose-fitting supports will solve this problem satisfactorily. Plastic must be protected, like copper, from mechanical damage such as errant nails and heavy objects.

Pipe and Fittings

Plastic pipe commonly used for plumbing is made out of four different kinds of plastic: PVC (polyvinyl chloride), CPVC (chlorinated polyvinyl chloride), ABS (acrylonitrile-butadine-styrene), and PB (polybutylene). PVC and CPVC are rigid, white or sometimes pastel plastics. CPVC, the newer of the two, is best when you want plastic hot-water pipes. It is, of course, also used for cold-water piping. ABS is rigid, black or gray plastic used primarily for drain, waste, and vent systems. PVC is also widely used for DWV systems. Most codes specify that pipe and fittings of different plastics (like ABS and PVC) cannot be mixed in the same system. The newest member of the group is PB pipe. It is flexible, black or dark gray, and more costly than the others. It cannot be joined with solvent cement but requires compression fittings. PB, now used mostly in irrigation systems and by gas and water companies for service mains, is not well known or readily available to the homeowner. However, because of its flexibility and strength, it is rapidly gaining acceptance for residential use. PB has good heat resistance and is approved for hot-water lines.

Plastic pipe comes in all the same nominal diameters as iron and copper pipe, and there are adapter fittings to connect plastic pipe to a metal system. In addition to the adapters, plastic pipe has all the same fittings as metal pipe (see page 27).

Plastic Fittings

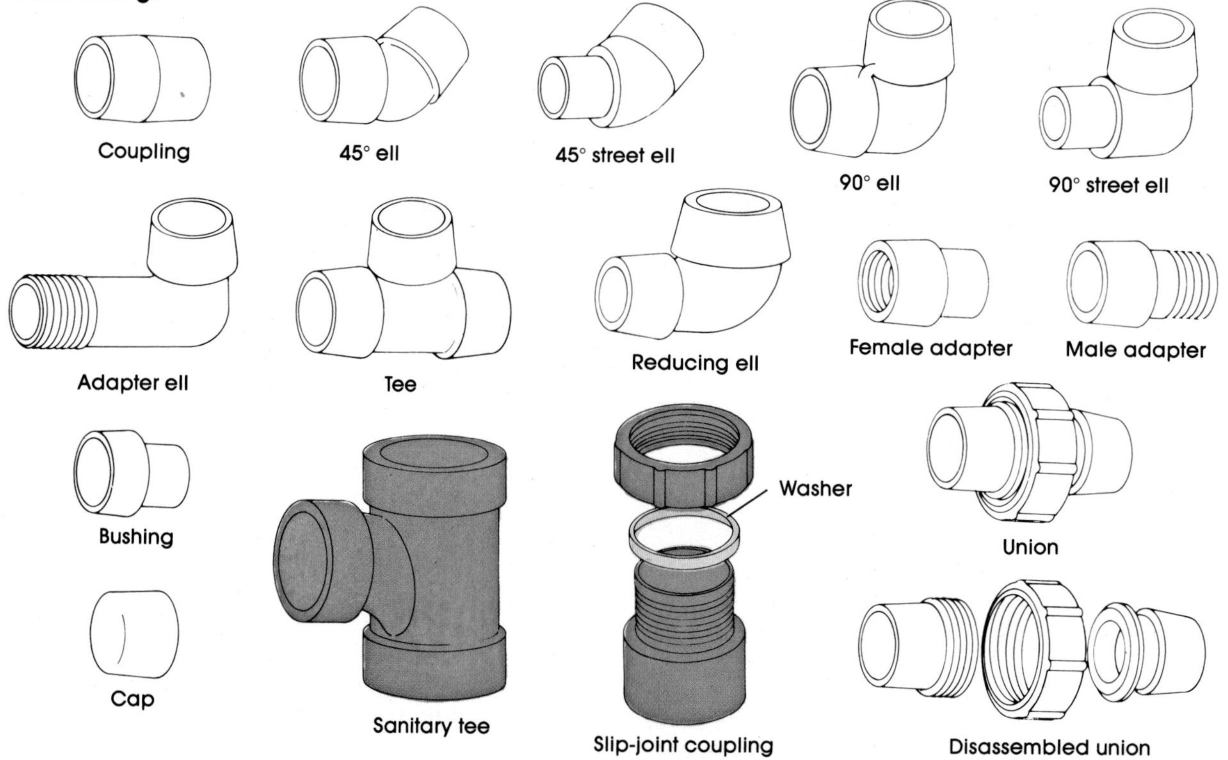

Cutting and Joining Plastic Pipe

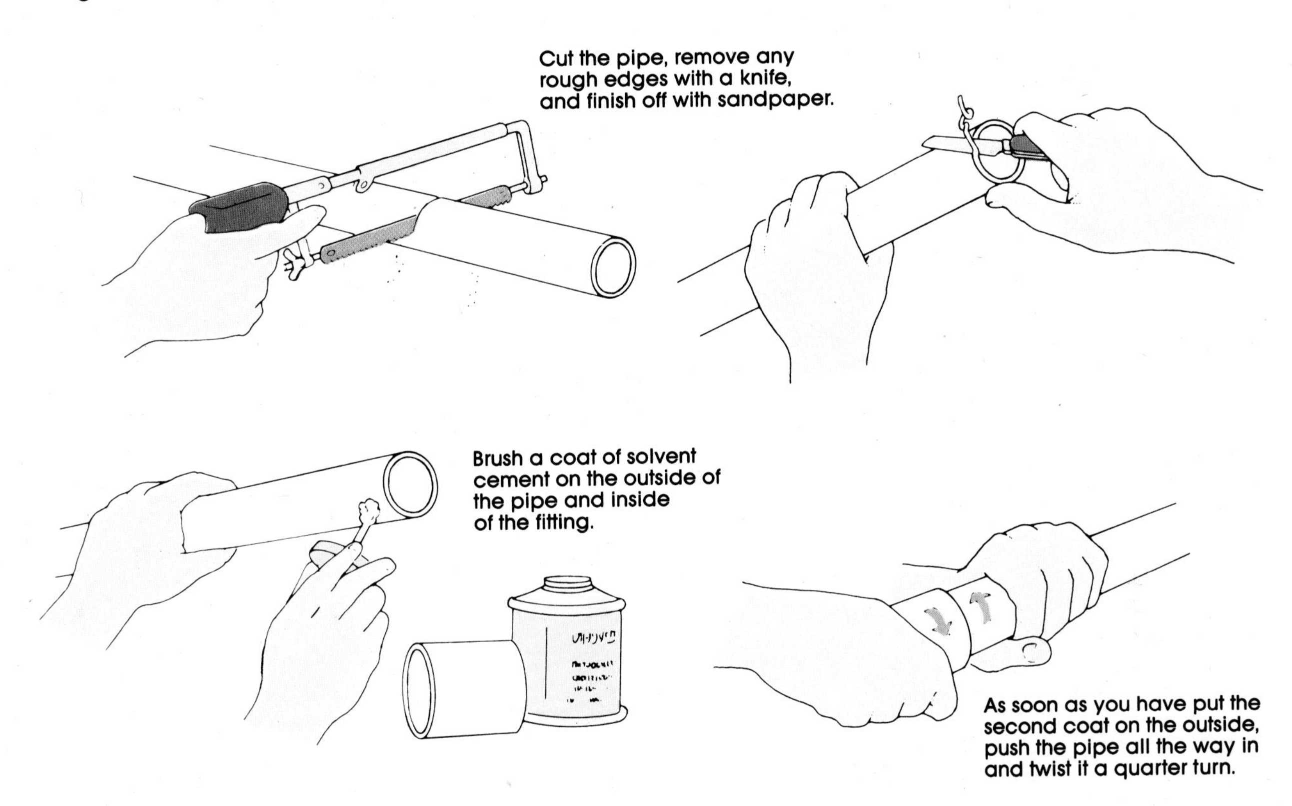

Measuring, Cutting, and Joining

Plastic pipe is measured and cut the same way as copper pipe (see page 33). You measure the distance between the faces of the fittings and add the length of pipe that will go into the fittings. Plastic pipe is so light and easy to maneuver that you can often just cut and slip together several sections of pipe and fittings to be sure it fits the way you want it, and then go back and cement it all together at once.

The actual cutting is done just like with copper tubing. But the cutting wheel for copper tubing is usually not sharp enough and just creases the plastic. You can use a tubing cutter that has been adapted for plastic pipe. If you want to use a saw, a hacksaw is probably best, but a coarser blade works better on plastic than the 32-teeth-per-inch recommended for copper tubing. In fact, almost any wood saw works quite well on plastic pipe. Be sure to remove burrs and rough edges with a knife or file before joining the pipe.

To join plastic pipe, you must use the solvent cement made for the kind of plastic you are using. ABS solvent will not work well on PVC pipe, for instance. The proper solvent actually melts the plastic of the pipe and fitting and "welds" them together, making the joints very solid and permanent.

Solvent cements are classified as both airborne contaminants and flammable liquids. Be sure to use them only in a well-ventilated area. If you must join pipe in an enclosed space, use a fan to provide ventilation and take frequent breaks to breathe clean air. Keep the container tightly closed when you're not actually using the solvent and keep the solvent and its container away from all sources of ignition, heat, sparks, and open flame.

When the pipe is all cut and ready to join, inspect both the end of the pipe and the fitting to be sure they are smooth and perfect with no deep cuts, cracks, or rough edges. Use some fine sandpaper or some specially prepared plastic pipe cleaner to remove the gloss and any moisture or oil from the joint surfaces. Brush a light coat of solvent cement on the outside of the pipe and inside the fitting. Then quickly brush a second coat on the outside of the pipe, push the pipe into the fitting as far as it will go, and give it a quarter turn to evenly distribute the solvent. A small bead of solvent should be visible all around the joint.

If the direction of the fitting on the pipe is critical, mark both pieces with a pencil or small knife cut before you apply the solvent (see the drawing). Line up the marks quickly after giving the pipe its quarter turn. Once the plastic hardens, it is a permanent joint and cannot be adjusted.

Hold the joint together for 10 or 15 seconds to be sure it is set solidly before you move on to the next joint. Wait at least an hour after you've made the last joint before you test the system. In fact, it's better to let the plastic set overnight before you turn on the water.

Fixing Leaks in Plastic Pipe

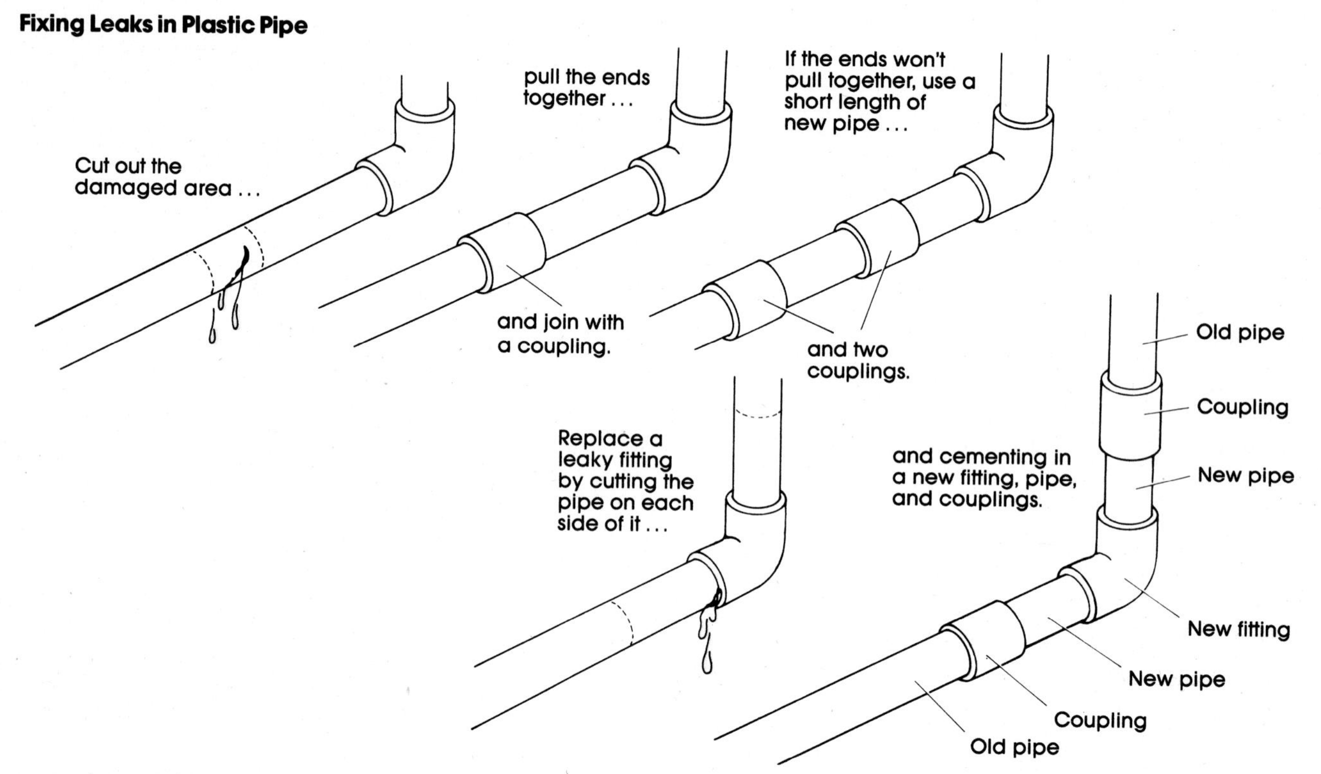

Repairing Plastic Pipe

For a pinhole or small leak, turn off the water, drain the pipe, and let it dry for a few hours. Force some plastic solvent cement into the hole and wrap the area tightly with plastic electrical tape.

For a larger hole or damaged pipe, cut out the bad section and if there is enough play in the system, rejoin the pipes with a coupling. If there isn't any play, use two couplings or a coupling and a plastic union to add a section of pipe the same length as the piece that you cut out.

If you have a leaky joint, you can try smearing or forcing solvent cement into the joint, but this doesn't usually work. The best solution is to cut the pipe a couple of inches on each side of the joint and install a new joint and short sections of pipe with couplings (see the drawing).

Tools for Working with Plastic Pipe

Tubing cutter or saw. A tubing cutter designed for plastic pipe works best. The cutting wheel designed for copper tubing will not cut the plastic satisfactorily. Almost any saw will cut plastic pipe. Those with fine teeth, like a hacksaw, work best. When cutting plastic pipe with a saw, be sure that the cut is at a precise right angle. A small miter box is useful for this purpose. Also be sure to clean off all the burrs and smooth the edges of the plastic with a knife or sandpaper before joining.

Solvent cement. You must use the proper solvent cement for each pipe—for example, PVC cement for PVC pipe and ABS cement for ABS pipe. The cement usually comes with a brush or dobber attached to the lid of the container for easy application.

Plastic Pipe Tools

CAST IRON PIPE

Although plastic pipe or large copper tubing with soldered joints is used a lot now for home drain, waste, and vent systems, there is still a place for traditional cast iron pipe. Older homes, of course, have cast iron DWV systems that occasionally need repair. And some codes say that because of its strength, cast iron should still be used for underground installations where heavy pressures would crush alternative pipe. Even though cast iron is very heavy and inconvenient to install and repair, it is also very durable and trouble-free once it's in place.

Pipe and Fittings

Cast iron pipe comes in 5- and 10-foot lengths and in diameters of 2 inches and larger; 1½-inch hubless pipe is now available in some areas. For home use, 2-, 3-, and 4-inch pipe is the most common. The pipe and fittings come in two types. The older type, joined with oakum and molten lead, is called hub or hub-and-spigot pipe. The newer hubless pipe and fittings are joined with special clamps and a rubber sleeve.

Hub pipe usually has a hub on one end and a spigot on the other. The spigot of one section fits into the hub on the next. You can buy double-hub sections for cutting when you need pieces shorter than the standard 5- and 10-foot sections. When you cut the length you need from a double-hub section, it leaves you another usable piece instead of a long piece of wasted pipe. Of course, the ideal way to cut down on waste of both time and pipe is very careful planning. If you are planning a new installation (or even a slightly complicated repair), draw it out carefully and try to figure out how to do it with standard lengths as much as possible.

An installation of hubless pipe automatically avoids this kind of waste because all of the pipe and fittings have smooth, even ends. The ends just butt together inside a neoprene sleeve that is held tightly in place by a stainless steel clamp. It is becoming rare now, but building or plumbing codes in some localities still prohibit the use of hubless DWV pipe. Be sure to check before you use it.

Both hub and hubless fittings come in a seemingly endless variety. Elbows are called *bends* in the DWV system. They come in several more angles than water supply pipe does. The 1/16-, 1/8-, 1/6-, 1/5-, and 1/4-inch bends are at angles of 22½, 45, 60, 72, and 90 degrees respectively. They also come in short, regular, and long, which indicates the sweep or radius of their curve. Closet bends are special fittings designed to connect the toilet or water closet to the drain system. These are made in several

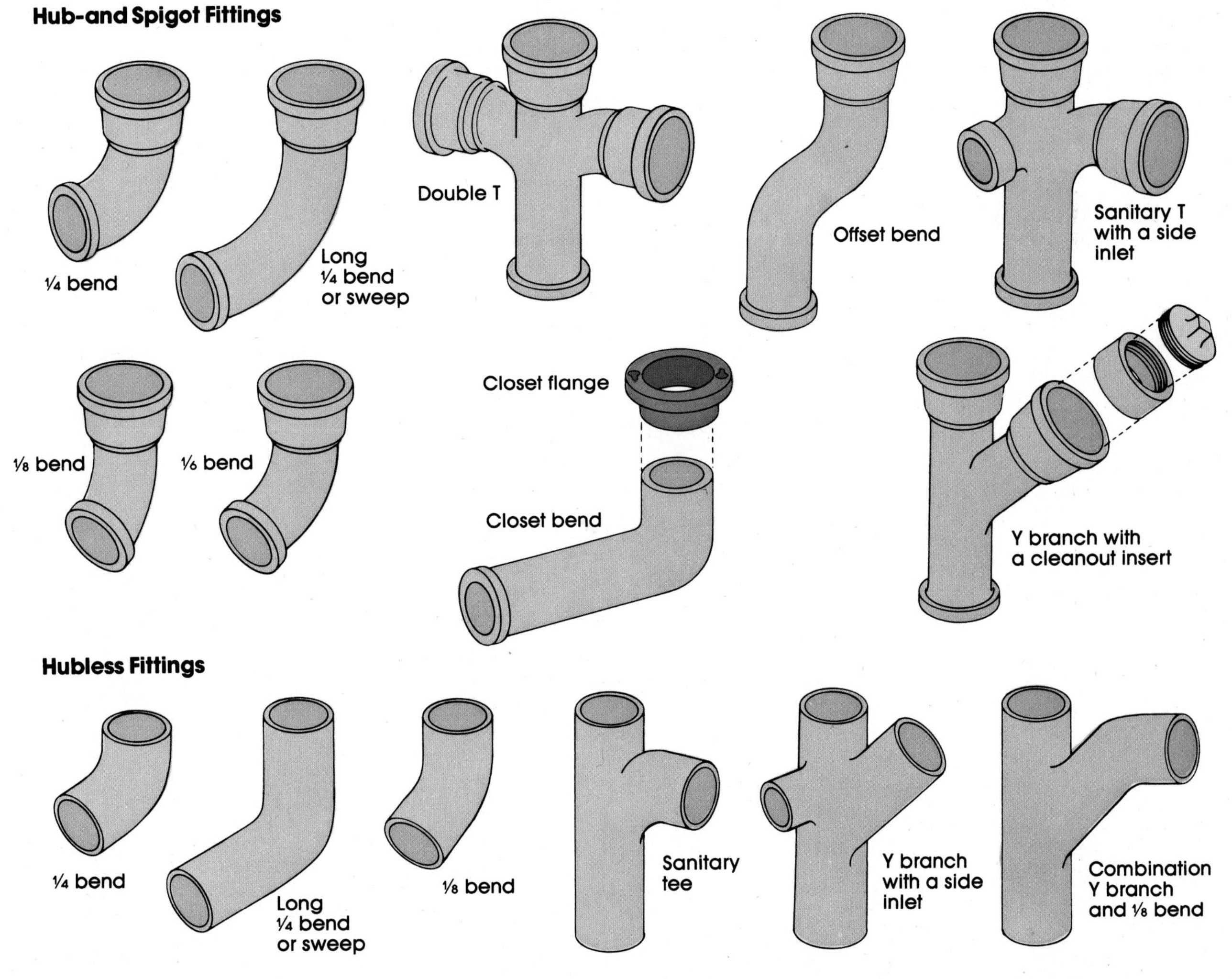

styles; those required by your local codes are usually stocked by the dealers in your area. There are also reducers, increasers, and offset bends. A special flange called a closet flange goes between the toilet and the closet bend.

Tees and wyes come single or double, with the branches on the same plane or to the side and with single and double side inlets for the connection of smaller drains or vent pipes. The tees and wyes come in two types: *sanitary* fittings designed to carry waste material and *straight* fittings to carry gas only in the vent part of the system. Cleanout or test tees have one branch threaded with a plug. These are used by the inspector to test the system and as permanent cleanouts.

Measuring and Cutting

Hub pipe and fittings are measured from hub face to hub face. Then you add the amount of pipe that goes into the fitting to get the total length of the pipe you'll need. Hubless pipe and fittings are measured end to end, less the width of the separator in the neoprene sleeve, usually about 1/8 inch. Mark the place to cut with chalk or crayon. Draw a line all the way around the pipe to be sure you cut it off square.

Both hub and hubless cast iron pipe are cut with the same methods. If you are repairing a leak or adding to a system and need to cut a pipe that is already installed, you'll need a pipe cutter. It is called a *soil pipe cutter* and can usually be rented.

Cutting Cast Iron Pipe

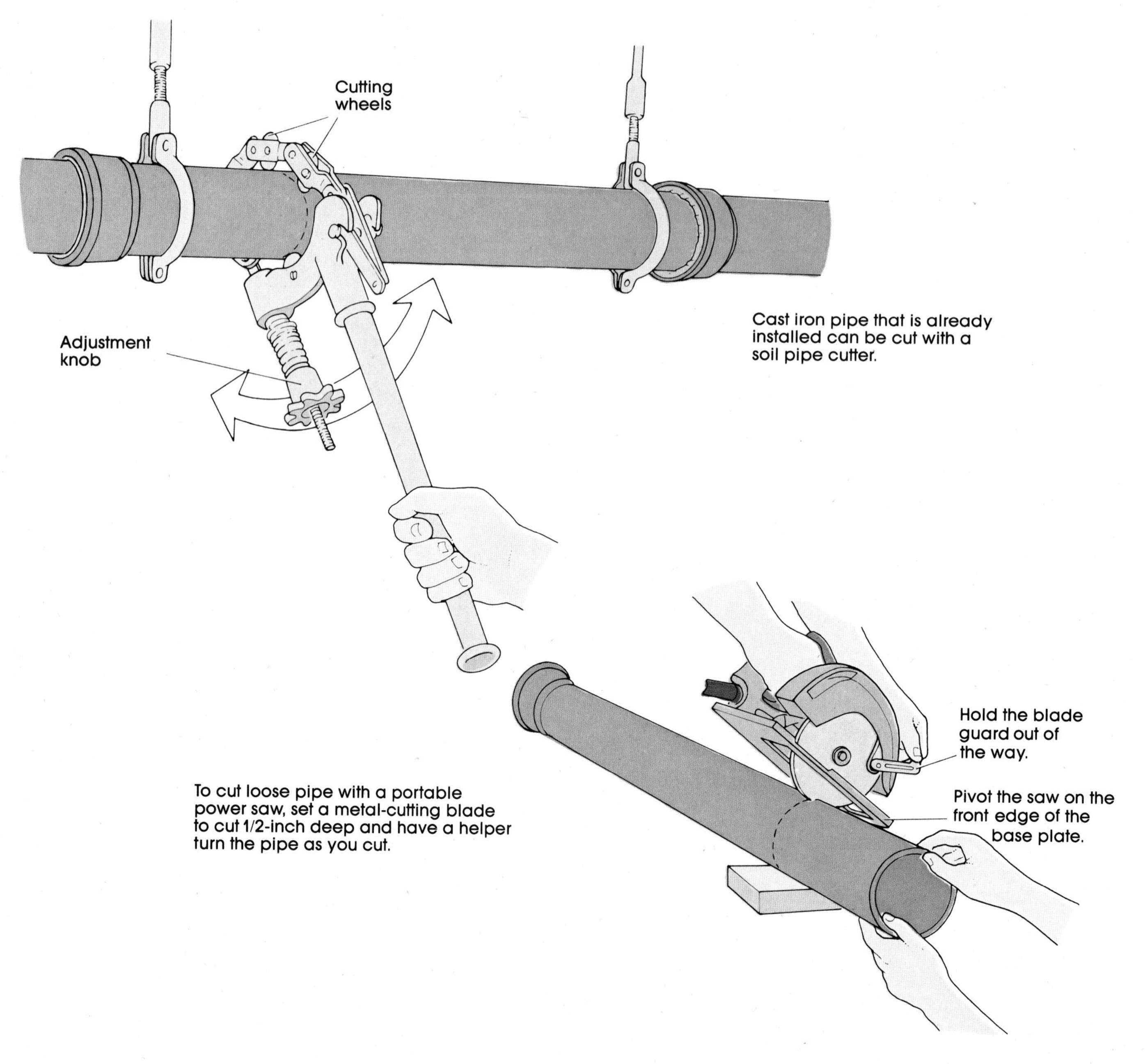

Cast iron pipe that is already installed can be cut with a soil pipe cutter.

To cut loose pipe with a portable power saw, set a metal-cutting blade to cut 1/2-inch deep and have a helper turn the pipe as you cut.

A soil pipe cutter works much like the pipe cutter or tubing cutter described on page 29. It has a pipe handle and two or more steel cutting wheels. The pressure is put on the cutting wheels by turning a knob. You tighten the knob and rotate the cutter around the pipe alternately until the pipe snaps off.

If you need to cut loose, uninstalled pipe there are several ways. One is to put the pipe in a vise and use the soil pipe cutter. Other methods involve a portable power saw, hacksaw, and hammer and chisel.

To use the power saw, you'll need a metal-cutting blade and a helper. Set the blade so it cuts about ½-inch deep. Have your helper turn the pipe slowly as you follow the chalk line with the saw.

You can also simply use a ¾-inch cold chisel and a 12-ounce ball peen hammer. Lay the pipe on boards or a mound of dirt so it is well supported. Hold the chisel on the chalk line and tap it lightly. Move around the pipe until it is scored all the way around. Then go around again and again using slightly harder blows for each revolution. The pipe will usually snap off about the third or fourth time you go around. Go slowly until you get a feel for what you are doing. If you hit too hard, the pipe may crack somewhere besides on your chalk line. The hammering method can be speeded up by scoring the pipe first with a hacksaw. Make a cut about 1/16-inch deep all around the cut line. Then go around the pipe once or twice, tapping on the saw cut with the hammer.

(Cutting cast iron pipe continued)

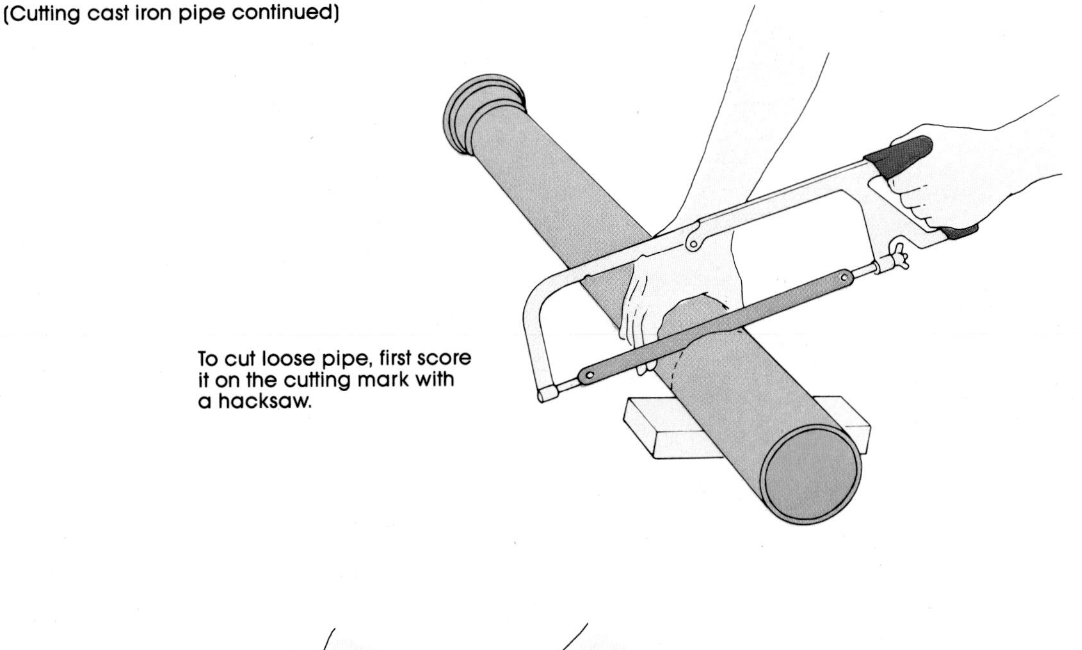

To cut loose pipe, first score it on the cutting mark with a hacksaw.

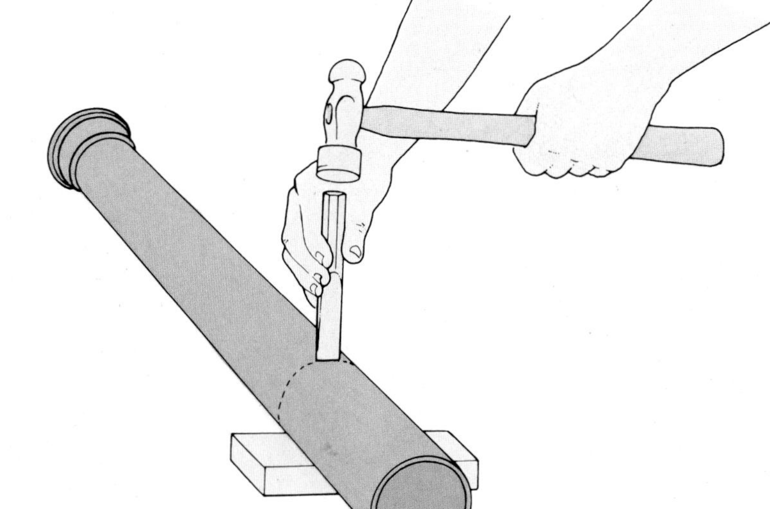

Then hold a chisel on the scored line and tap it lightly. Move around the pipe three or four times until it snaps off.

Joining

The traditional way to join hub pipe is with a plumber's furnace or some other way of melting lead, a ladle, a caulking tool, a joint runner, and some oakum and lead. This potentially dangerous method can be avoided altogether by using lead wool instead of molten lead. If you use the lead wool method, insert the spigot end of one piece of pipe into the hub of the next piece. Use the caulking tool to pack oakum into the joint until it is about one inch from the face of the hub. Then fill the remaining space with lead wool that you pack tightly with a caulking tool.

If you opt for the molten lead method, follow these instructions. If the pipe is vertical so the hub is straight up, you won't need the joint runner. If the hub is tipped sideways, secure the runner around the joint with its clip. Have the opening of the runner at the highest point on the rim of the hub.

After packing the joint, melt the lead, dip it with the ladle, and pour it into the joint until the hub is full to the

Joining Cast Iron Pipe

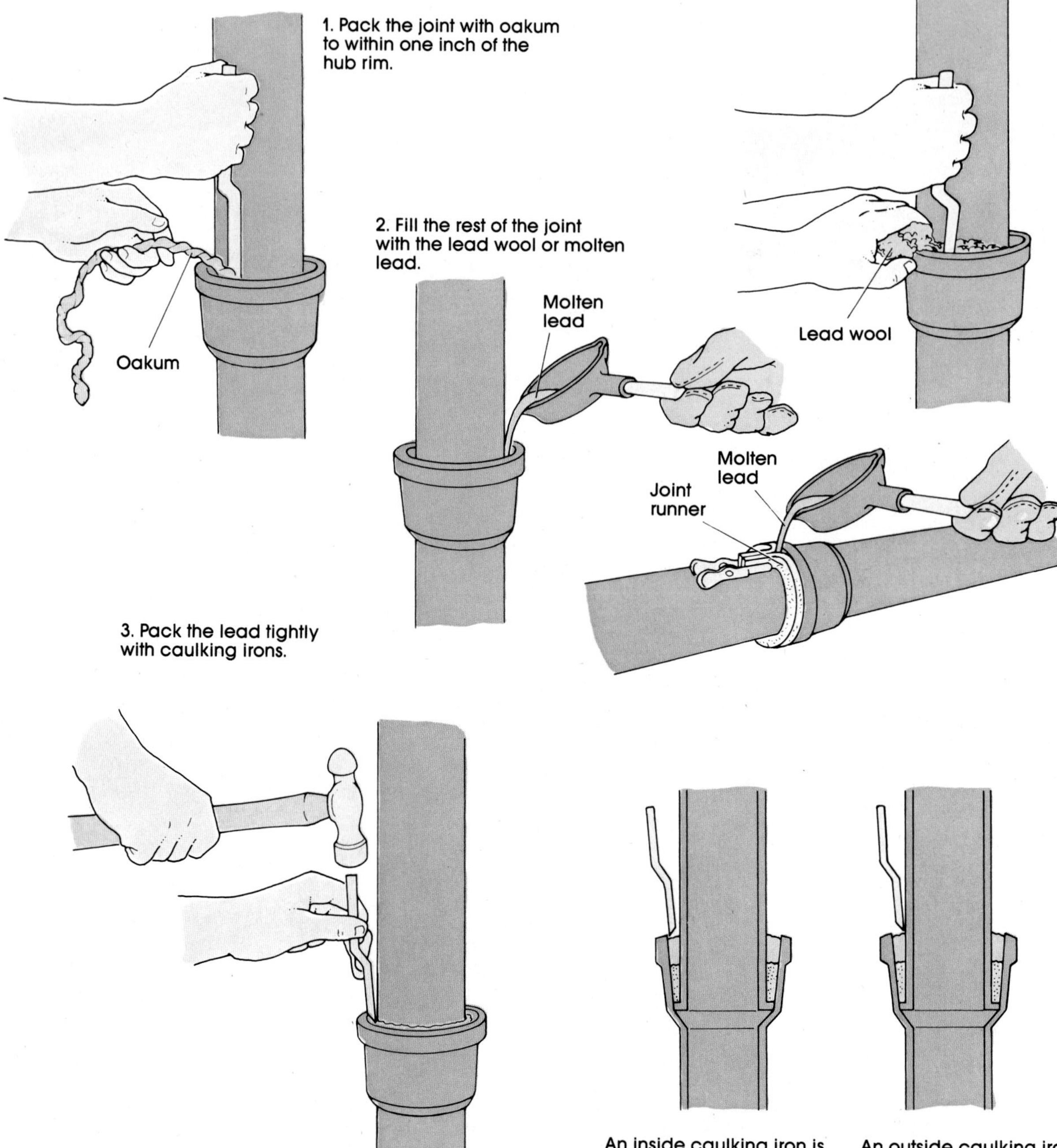

An inside caulking iron is beveled to pack the lead against the spigot.

An outside caulking iron is beveled to pack the lead against the hub.

rim. Let the lead cool and then use the caulking iron to pack it down firmly. With either the lead wool or molten lead method, you should have two caulking irons—an outside and an inside iron. Use them alternately to pack the lead against the hub and the spigot.

■ WARNING: The slightest bit of moisture, when hit by molten lead, turns instantly to steam. It expands quickly—explodes actually—and can throw molten lead all around. Be sure that your tools are dry and there is no moisture in the joint or oakum when you pour in the lead. If you should suspect that there might be moisture in the joint, heat it with a propane torch to dry it before you pour the lead. Wear safety glasses or goggles.

To join hubless pipe, you'll need a sleeve and clamp for each joint. These are sold in sets. The only tool you'll need is a stout screwdriver. Place the neoprene sleeve over the end of one pipe. Slip the clamp onto the other pipe and slide the pipe into the neoprene sleeve. Center the clamp over the sleeve and tighten the screws on the clamp. That's all there is to it.

Joining Hubless Pipe

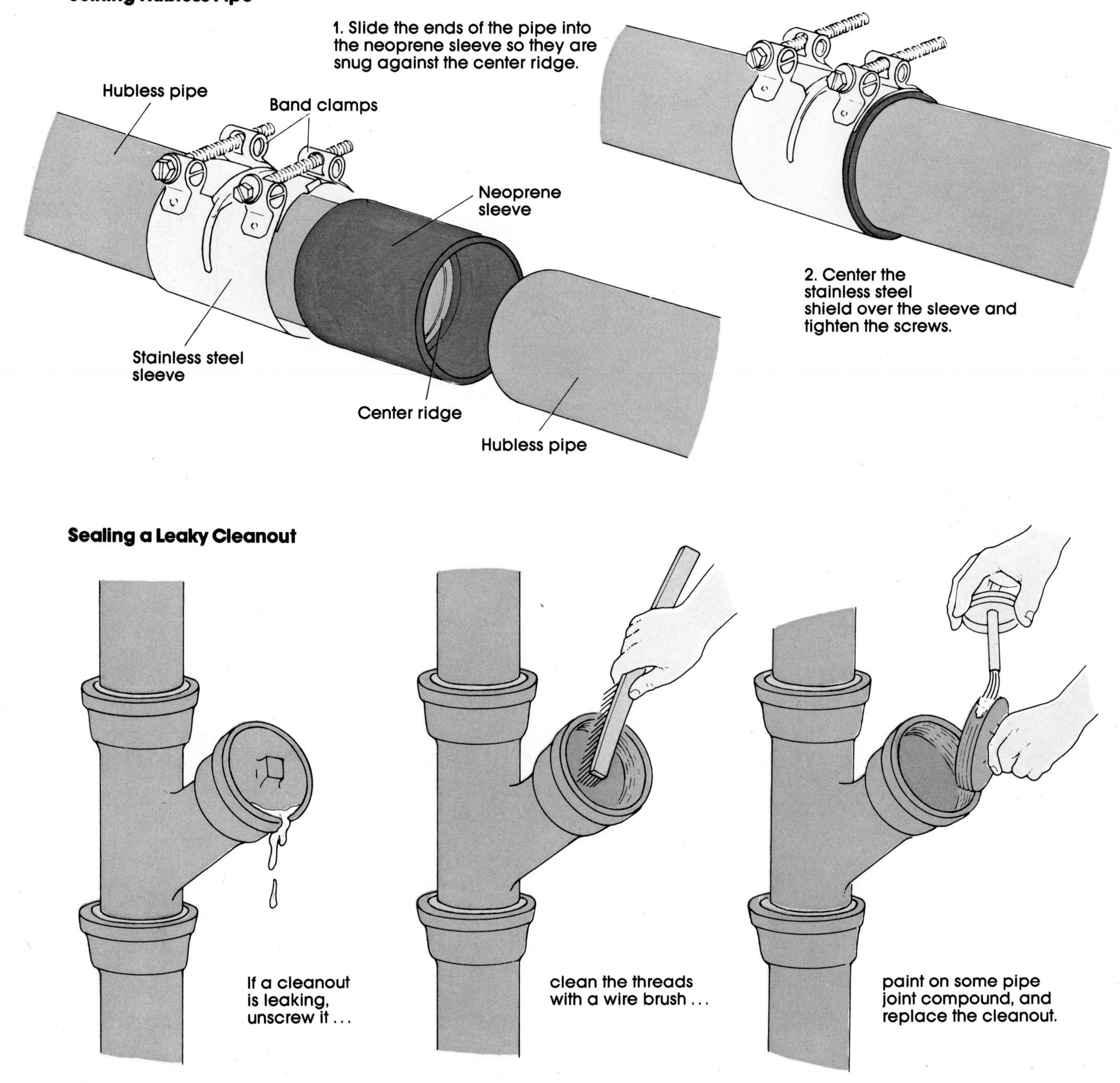

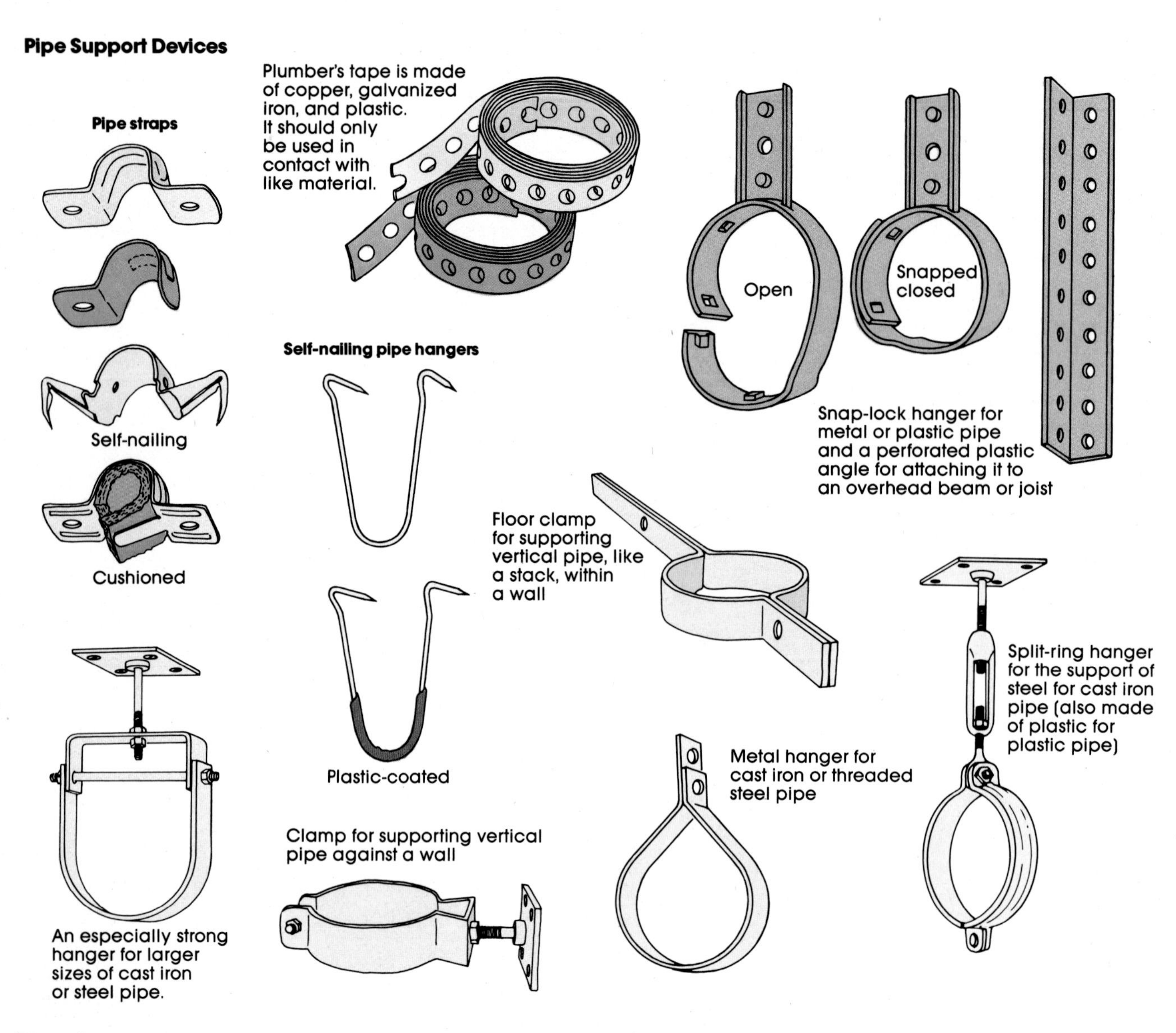

Pipe Supports

Horizontal runs of DWV pipe must be supported with approved hangers at designated intervals according to building and plumbing codes. The spacing for supports varies with the kind of pipe. Here are the usual standards. Check your local code for specifics.

Hub cast iron pipe must be supported at intervals no greater than 5 feet. Hubless pipe must have a support at every joint. Threaded iron pipe must have a support every 10 feet or less. Soldered copper pipe and plastic cold-water pipe must have a support at 8-foot and 4-foot intervals respectively. Plastic hot-water piping needs continuous support.

The interval between supply pipe supports is not usually spelled out in the codes, but they must be secured to the building structure to minimize the movement of the pipe from hammering (sudden starting or stopping of the water in the pipe) and from the expansion and contraction of the piping itself due to temperature fluctuations.

There are many manufactured bulk materials and gadgets to support and secure pipe. Plumber's tape, which comes in 10-foot rolls, is probably the most universally available material. It is pliable, galvanized iron or copper-plated strapping. It is perforated along its entire length so nails, screws, or bolts can be easily stuck through it. Do not use this metal support material for plastic pipe.

Other hangers, straps, brackets, and supports vary in form from manufacturer to manufacturer and from store to store. A sampling of these is shown in the drawing.

Condensed moisture or dampness between unlike metals causes galvanic action even on the outside of pipes. Therefore, you should always try to match the metal of the support with the metal of the pipe. Use galvanized iron or steel supports for iron pipe and copper or brass supports for copper or brass pipe. There are also many hooks, hangers, and other supports for pipe and tubing that are coated with a rubberlike plastic or have a felt-like fiber insert. These can be used freely on all kinds of pipe. Since the diameters are similar, most plastic-coated devices made to support copper tubing can also be used to hold plastic piping.

Tools for Working with Cast Iron Pipe

Hacksaw, ball peen hammer, and cold chisel. This is probably the best choice of methods to cut cast iron pipe before it is installed (see page 43).

Portable power saw with a metal-cutting blade. To cut cast iron pipe with a portable power saw, you'll need a metal-cutting abrasive-wheel blade, a pair of safety glasses, and a helper (see page 43).

Soil pipe cutter. This tool is absolutely necessary for cutting cast iron pipe that is already in place. It is probably not worth your while to buy one, as they can be rented at most equipment rental outlets. If the pipe you want to cut is unattached, one of the other methods of cutting is probably better.

Plumber's furnace and ladle. Here is another item you can rent for an occasional job. It is used to melt the lead to join hub-and-spigot sewer pipe.

Joint runner. This device holds molten lead in the joint when you join horizontal sections of hub-and-spigot pipe.

Caulking tools. There are several tools used to pack the oakum and lead in to the joint of hub-and-spigot pipe. They come both separately and in combination—that is, a different tool on each end of a single handle. The inside caulking iron and outside caulking iron will be all you'll need for most jobs. The yarning iron, packing iron, and pickout tool are all variations of the caulking irons and will make this job a little easier if you are making very many joints.

Oakum and lead or lead wool. Oakum is the yarnlike material that, when packed tightly and exposed to moisture, expands to seal joints or cracks. The lead in a hub-and-spigot joint gives the oakum a firm backing against which to expand.

No-hub fittings. These are neoprene sleeves and stainless steel clamps designed to join hubless cast iron sewer pipe. These kinds of joints are easier and quicker to make and probably just as durable as hub-and-spigot, oakum-and-lead joints.

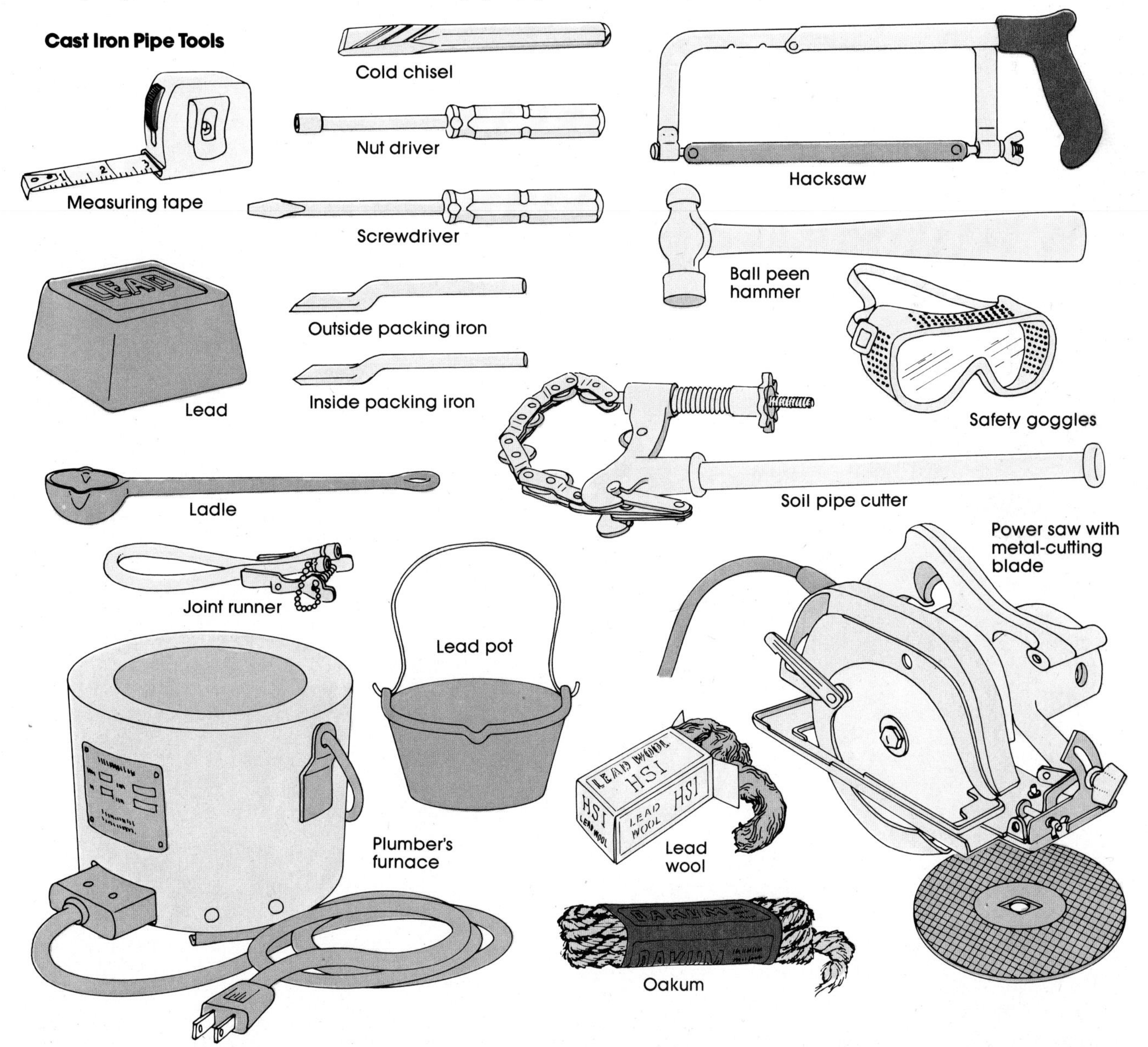

125W
1
½
150

VALVES & FAUCETS

You don't need to let that faucet drip another night. What you should know about repairing and replacing valves and faucets is all here. And installing new ones may be easier than you think.

A valve is a device that regulates the flow of liquid or gas within, entering, or leaving a system of pipes. It can stop, start, increase, decrease, or change the direction of the flow. A faucet is a valve at the end of a pipe that regulates the flow from the system. Plumbers almost never refer to a faucet as a valve, even though it is one.

Valves are usually named for either how they work or for the function they perform. The descriptions and drawings of the valves and faucets in this book are very general. In reality, each manufacturer designs and makes a unique product. The parts cannot generally be interchanged from one to another. Even when the products of one manufacturer change from one year to the next, the parts from the new model won't necessarily interchange with those from their earlier one.

Here are the common types of valves used in home plumbing systems. Note that water and gas valves are not interchangeable. Use only gas valves for gas.

If you want to install more modern faucets, you will find that a lot has been standardized.

Gate valve. This valve has a tapered disc or wedge-shaped "gate" that moves across the opening to stop the flow of water and pulls out of the way to let the water flow. When this valve is open, the water can flow straight through with no obstruction or change in direction to impede its full pressure, velocity, and volume. Gate valves are used where a line is either fully open or fully closed most of the time.

Globe valve. This is a compression-type valve with a disc and seat that are set in a partition parallel to the flow of the water. When the valve is open, the water must change direction twice in order to get through, which reduces the pressure, velocity, and volume. This is a poor choice of valve for any high-volume water supply line that usually remains fully open. Globe valves are used where a line is opened and closed frequently or where a valve is needed to control the volume and the loss of pressure through the valve will not be problematic.

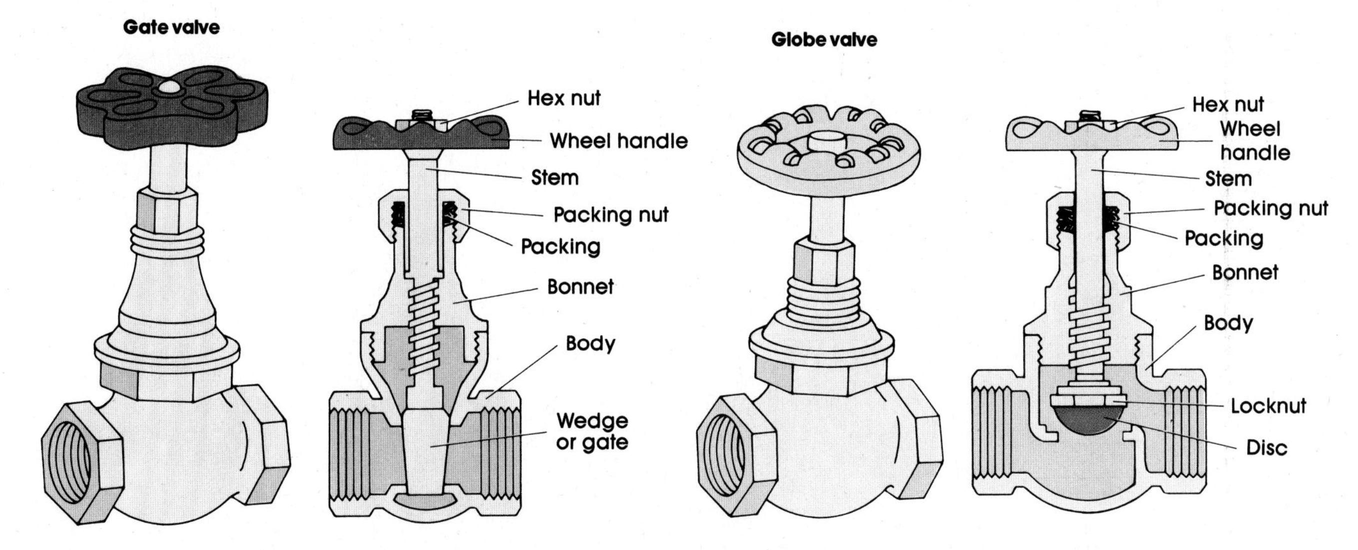

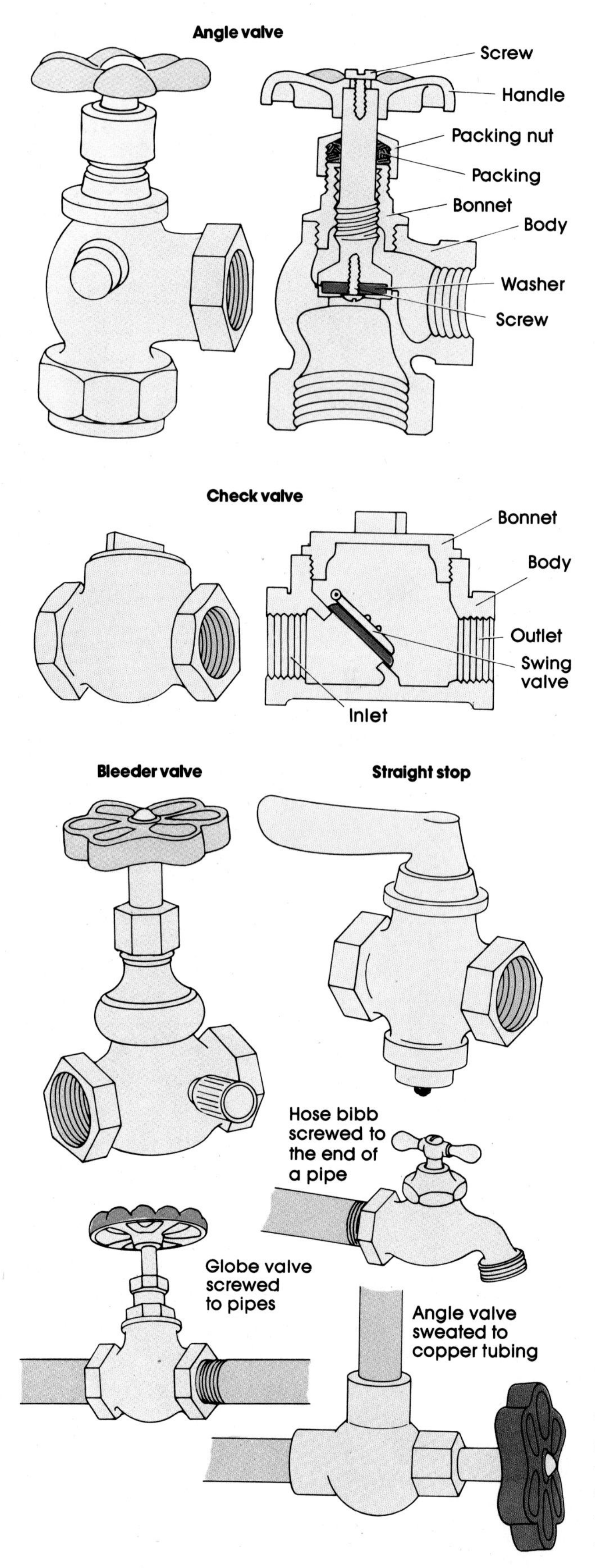

Angle valve. The mechanism in this valve is very similar to that in a globe valve, but with openings at right angles to each other like an elbow. Because the water only changes direction once instead of twice, an angle valve impedes water flow less than a globe valve. By inserting an angle valve where a pipe turns, you get the benefits of a globe valve without its main drawback, and you eliminate the need for an elbow.

Bleeder valve. Also called a *stop and waste* or *drainable* valve, it has a plug on the outlet side that allows water to be drained from the system when the valve has the water shut off. This kind of valve or some other way of draining the system is required by code in cold-weather areas where pipes can freeze in the winter.

Check valve. This automatic valve allows the free flow of water in one direction, but closes if the liquid starts to flow the other way. The primary use of a check valve is on private water supply systems between the well and the pump. It keeps the water in the system from draining back into the well when the pump stops.

Straight stop. This valve is usually made of brass and is used for gas. It has a tapered ground plug that fits in a tapered ground body. The plug is mounted on a spring so the tension keeps the plug securely in place and gastight. When the plug is turned so the hole through it is parallel to the gas flow, the gas goes right through. Turning the plug so the hole is perpendicular to the flow stops the gas. When used as a supply stop near an appliance, a straight stop must have a manually operated handle. For other applications, it is equipped with a square, hexagonal, or specially shaped head to be operated with a wrench or a special tool.

There are two basic types of faucet used in turning water on and off. One is a very old design called a *washer* or *compression faucet.* The more modern type, called a *noncompression* or *washerless faucet,* mixes both hot and cold water regulated by one knob or handle. Most new kitchens and bathrooms are equipped with washerless faucets.

A compression faucet closes by a screw pushing and compressing a washer against a seat. The washer wears out periodically and can be replaced quite easily (see page 52). Whenever there is a single faucet or a combination faucet where hot and cold water are turned on and off by separate handles, you can be pretty sure it is a compression or washer-type. A compression faucet with the handle on top and the spout turning downward is called a *bibcock* or *bibb* for short. If the spout is threaded so you can screw on a hose it is called a *hose bibb.*

Washerless faucets come in three kinds: valve, ball, and cartridge faucets. Because no manufacturer's parts are interchangeable with others or even with parts from another model by the same maker, these are difficult to repair if anything goes wrong with them. Often the model you have has long since been discontinued by the time something goes wrong with it. Washerless faucets are usually much more expensive than their counterparts with washers. However, the new washerless types are very trouble-free and can often be used for many years without probelms.

Types of Faucets

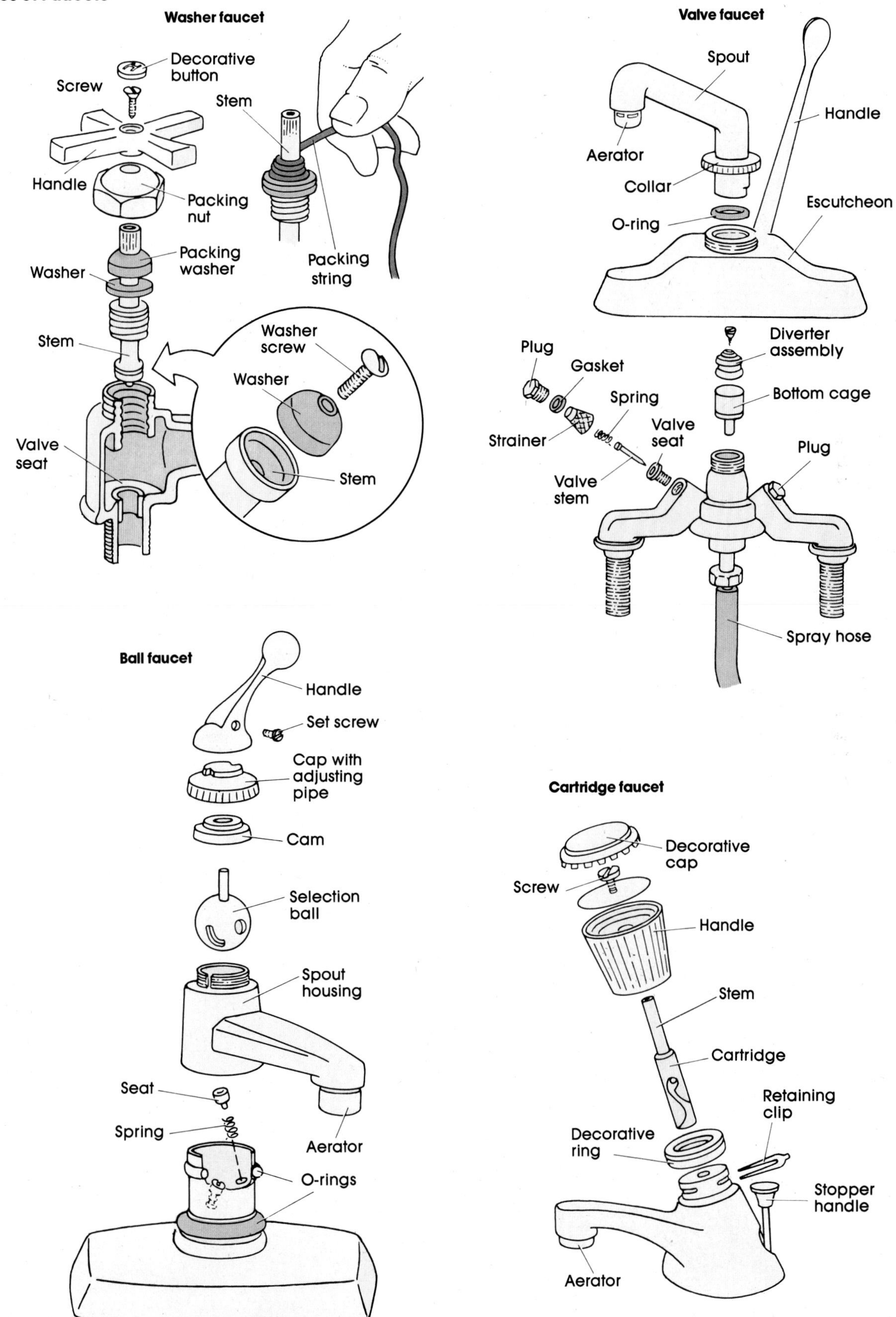

REPAIRING FAUCETS

Repairing the washer-type or compression faucet is a fairly straightforward process. But if you are having trouble with a noncompression type, you may have difficulty finding replacement parts. Because there is absolutely no standardization among the many different varieties on the market, you must buy parts for the same make and model that you have. If your particular model has been off the market for a while, parts may be impossible to find and you will be better off just buying a new faucet.

The faucets that control the water in tubs and showers are essentially the same as those on sinks. Most of them are the compression type, but those which operate with one handle are usually the cartridge type. Both are disassembled and repaired in the same manner as described for sink faucets (see the drawings).

Compression Faucet Repair

If your faucet doesn't seem to be putting out as much water as it used to, the problem just might be a clogged aerator on the nozzle of the faucet. Unscrew the aerator from the nozzle, take the pieces apart, and wash them clean under the faucet. Notice how the pieces fit so you can put it back together the same way it was. If you can't unscrew the aerator by hand, wrap some adhesive or electrician's tape around it to protect its finish, and turn it counterclockwise with a pair of pliers.

Leaking spouts. If a spout leaks when the water is turned off, you probably have a bad washer or seat that allows water to slip past. The repair is made through the handles. With two handles and one spout, you can tell which handle has the problem by turning off one of the supply

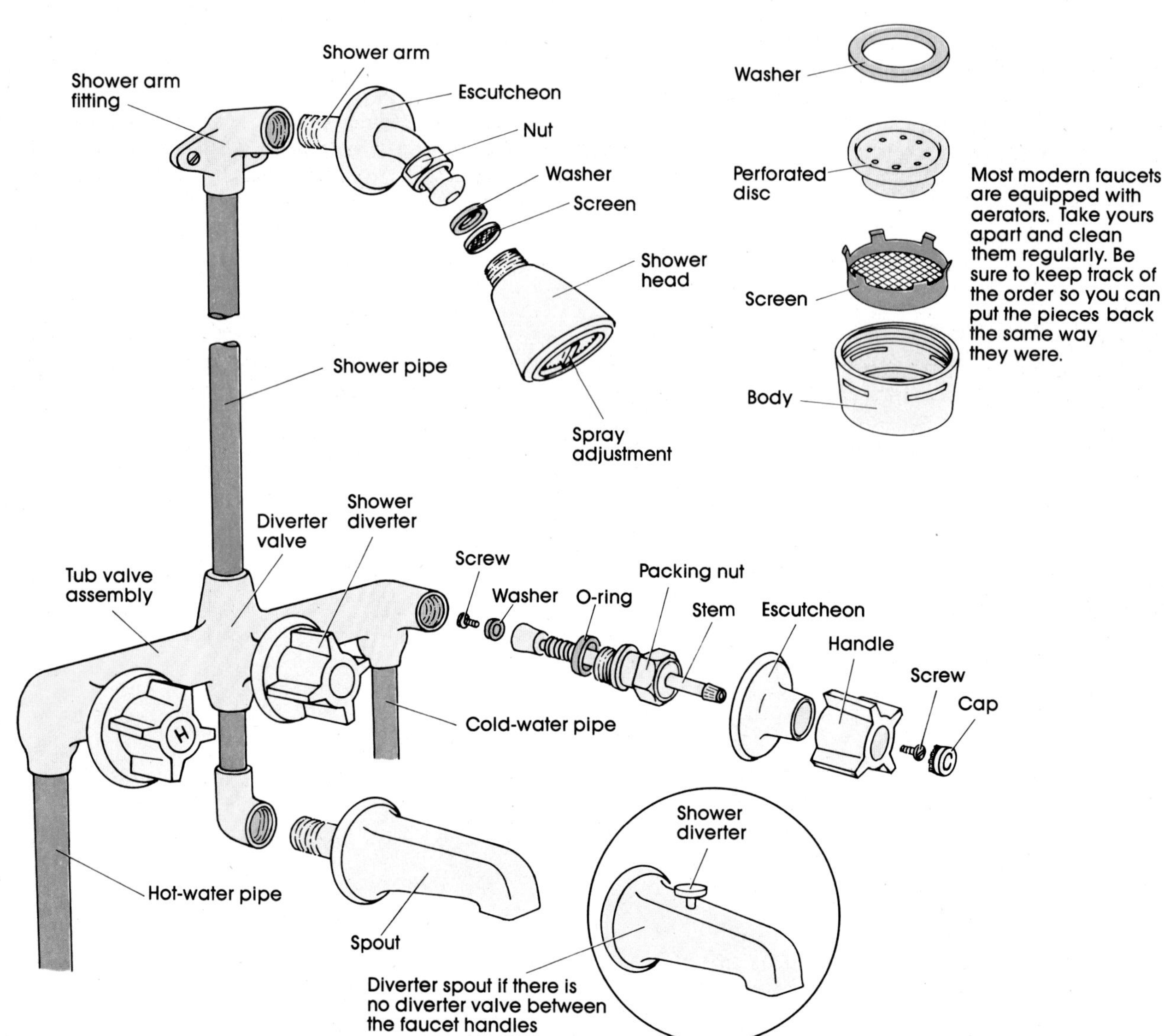

stops under the sink. If the leak stops when you turn off one of the stops, you have found which side has the leak.

The screw for removing the handle is right on top of the handle. However, it may be hidden beneath a decorative cap. If so, unscrew or pry off the cap before you remove the screw that holds the handle. Then pull off the handle. You may have to work it back and forth a little to get it loose.

Beneath the handle is the packing nut. Remove it by turning the nut counterclockwise with a wrench. Protect its chrome or gold-colored finish with tape. Then lift out the entire assembly. You will have to turn it counterclockwise a few turns as you remove it. On the bottom of the assembly is a brass screw holding the washer in place. Remove the screw and the washer. Replace the washer with one exactly the same size and shape and replace the brass screw if it is bent or worn. If the washer has a beveled edge, the bevel always faces toward the screw head. While you have the washer assembly out, inspect the valve seat, too. If it is rough or chipped, it should be resurfaced or replaced. Otherwise it will just continue to wear out washers very quickly.

Valve seats. A gouged or roughened valve seat prevents the washer from fitting properly. To see if it can be replaced, look for a square, hexagonal, or round hole in the center of the seat. If it has a square or hexagonal hole, use a special seat wrench or an Allen wrench to unscrew it (see page 6). Take the old seat to the store and get a replacement that matches it exactly.

If the seat has a round hole, it cannot be replaced and will have to be refaced. For this job, buy an inexpensive valve seat tool. Professional refacers cost well over ten dollars, but you can usually find a tool designed for homeowners that sells for three or four dollars. Make sure the one you buy has a guide or uses the faucet's packing nut for a guide. With the faucet dismantled, set the guide into the top of the faucet or put the shaft through the packing nut and screw it onto the faucet. Turn the refacing tool down until it touches the seat. Work it back and forth to shave the seat smooth.

Leaking handles. If your faucet is leaking around the shaft of the handle, remove the handle (as described under leaking spouts) and then loosen the packing nut and lift it off. Under the packing nut on the stem, you will see either a rubber O-ring, a packing washer, or wrappings of packing string. If it is an O-ring, work it off with a screwdriver, and roll on a new one of the same size from your washer packet. For packing, buy a new packing washer or some graphite-impregnated or teflon packing string. Wind the string clockwise around the stem until it is thick enough to pack—usually five or six turns. Then tighten down the packing nut and replace the handle.

Noncompression Faucet Repair

With all three basic types—cartridge, valve, and ball—most manufacturers do their best to hide the screws and nuts that hold them together. Some are hidden under snap or screw caps on top of the handle; sometimes a nut is located at the base of the spout; or there may be a set screw located under the handle. On some cartridge types, you find the set screw by pushing the handle all the way back and looking under it.

Cartridge faucets. These come in two varieties: the metal sleeve with a screw holding it from the top and the newer ceramic disc.

Disassemble the metal sleeve style by removing the screw, then pushing a screwdriver down the hole to keep the stem in place while you pull off the handle and cover. Then unscrew the retaining nut and remove the spout. With the faucet body exposed, you can see two O-rings—one at the top and one at the bottom. These are probably the source of your leak. Remove the retaining clip at the top of the faucet body, lift it off the stem, and replace the rings. Reassemble the faucet by putting everything back in the reverse order.

To remove the ceramic disc cartridge, tilt the knob all the way back to expose the set screw, loosen the set screw, and lift the handle assembly free. Unscrew the screws on top of the brass collar beneath the handle assembly. Lift out the disc cartridge assembly and examine it for wear and cracks or other damage. If the ceramic disc is damaged, it will have to be replaced with an exact duplicate. If the machined surface on which the disc rides is damaged, it will have to be remachined at a machine shop or the whole faucet will have to be replaced.

Valve faucets. Unscrew the collar at the base of the spout and pull the collar and spout free. If your only problem is a leak around the base of the spout, replace the O-ring and reassemble the faucet. Otherwise, with the spout off, lift off the *escutcheon* (faucet body cover). When the escutcheon is removed, you will see a hexagonal plug on each side of the faucet. Beneath each plug will be a gasket, strainer, spring, stem, and valve. Under the valve will be a valve seat that will have to be removed with a hexagonal seat-removal tool (see page 6). Remove all these parts and examine them for wear or damage. Take any parts that need replacement to your local dealer to be sure the new parts match, and then reassemble the faucet in the reverse order. Before replacing the escutcheon, adjust the set screw at the base of the handle so its movement is smooth but firm.

Ball faucets. Loosen the set screw at the base of the handle and lift off the handle. Wrap the screw cap beneath the handle with tape to protect its finish and unscrew it with a pair of channel-lock pliers. When you lift the ball-and-cam assembly out, you will be able to see two rubber valve seats underneath. If your faucet has been leaking, these valve seats are no doubt the problem. Pull them out with needlenose pliers and replace them. If the ball is corroded or gouged, it will have to be replaced, too. If not, slip it back into place. A slot in the side of the ball must line up with the metal projection on one side of the housing. The plastic cam assembly then slips on, its tab lining up with the slot in the faucet body. Before putting the handle back on, turn on the water and open the faucet. If there is a slight leak around the stem, tighten the adjusting ring inside the cam assembly by putting a screwdriver in the slot and turning the ring.

NEW VALVES & FAUCETS

If you just don't like your old faucets, replace them with some new ones. It is really not much harder than fixing the ones you have—sometimes it is a lot easier. To replace your old faucets, you must find a new unit you like that will fit the holes in your sink. Measure the space between the center of the holes. Draw a diagram like the one shown here and take it to the store with you.

If your old sink's holes don't match any single-lever assembly you can find, buy some modern individual faucets as replacements. *Do not* try to adapt your old sink to a unit that doesn't fit. It is not worth even trying. Chrome-plated caps are available to cover any unneeded holes in your sink.

Replacing Faucets

To remove your old faucets, start by turning off the water at the supply stops beneath the sink. Use a basin wrench to loosen the upper and lower coupling nuts on the supply tubes and then remove the tubes. Remove the lock nuts and washers that hold the faucets to the sink and lift them off.

Next take the faucet assembly from its box and check the parts list to be sure all the parts are there before you throw the box away. Use the enclosed instruction sheet to install the assembly onto the sink properly. It usually involves placing a gasket on the sink, setting the faucet assembly on the gasket, and securing it with washers and nuts from underneath.

Next, hold the supply tube in position and figure out what bends, if any, it needs to enter the valve and faucet perfectly straight on. If you have installed the valves so they point straight up, the supply tubes usually need two bends so the ends are parallel but offset a little (see the drawing). When you get the bends right, mark the length you'll need and cut the tubes off with a tubing cutter.

Slide the coupling nuts and compression rings onto the tubes, set them into the couplings of the faucets and valves, and tighten them as described before. You will probably need a basin wrench to reach the nut that holds the tubes to the faucets. Turn the supply stops on with the faucets off to check for leaks. Then, with the aerator removed, turn the faucets on to flush any debris from the pipes.

Typical Hole Spacing on Sinks

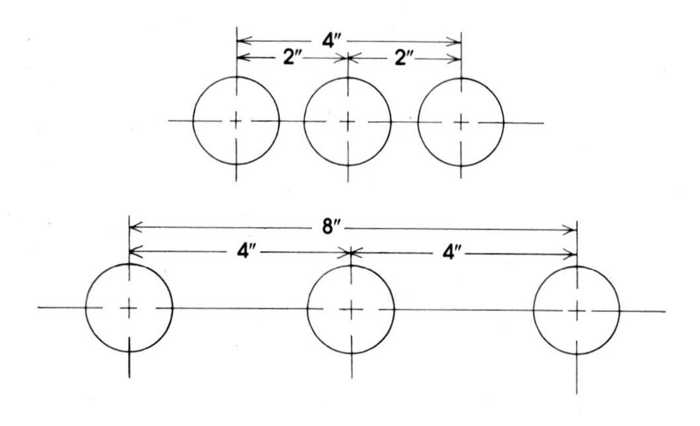

Installing Valves and Faucets

All of the valves listed on page 49–50 usually come with pipe-threaded openings and are screwed to the pipe just like any other fitting. If you have a soldered copper tubing system, you can get valves with solder-on, compression, or flare openings. Or you can sweat on a male threaded adapter and screw the valve to the adapter. Some plastic valves are equipped with nonthreaded solvent-welded openings, but most have threaded openings and are screwed onto an adapter that has been welded to the pipe.

Single faucets, like hose bibs, have a pipe thread base and can be screwed onto the end of any pipe. To install one in an existing pipeline, you can insert a tee into the line and then screw the faucet onto the tee.

There is a way to install a faucet on either a copper tube or iron pipeline without cutting the pipe. You can use either a *saddle-type faucet* or screw a regular faucet on a saddle-tee connector. A saddle connector clamps to the pipe with bolts, and the water enters the saddle and the faucet through a hole drilled in the pipe.

These days the entire installation of bathroom and kitchen faucets has become very standardized. Most faucets are hot and cold, single-handle faucets on kitchen and bathroom sinks. New sinks and almost all sinks that have been manufactured in the last 15 years or so, have a standard 4-inch or 8-inch space between the faucet holes to accommodate all standard faucet assemblies.

When the rough plumbing is done, stubs of the hot- and cold-water pipes are left protruding through the wall where the sink will be located. These pipes should be about 20 inches above the floor and about 8 to 10 inches apart. The hot-water pipe will be on the left when you face the wall. After the cabinets and sink are in place and secured, it is time to install the supply stops and faucets.

When you buy supply stops, you must be sure they are compatible with the pipe stubs. If you have threaded iron stubs, be sure the stop valves have standard IPS (iron pipe size) threads on the inlet opening. Copper tubing stubs give you a choice of the kind of fittings you want. You can get female solder fittings to sweat on, adapters to sweat on that will take a valve with pipe thread, and valves that have flare or compression fittings. Compression joints are much easier to install in this kind of situation, when you will be lying under a sink with your feet up the wall.

In addition to two supply stops, you will need the faucet assembly of your choice and two flexible supply tubes with compression fittings. Your hardware or plumbing dealer will help you pick out all the fittings you'll need to go with the faucet assembly you've chosen.

Because you are usually working with chrome-plated compression fittings in this kind of an installation, the main tools you'll need are a couple of smooth-jawed adjustable wrenches. A basin wrench is also desirable to tighten the nuts that are way up under the sink.

The first step is to turn off the water. You may be able

Saddle Tee Installation

Saddle Faucet Installation

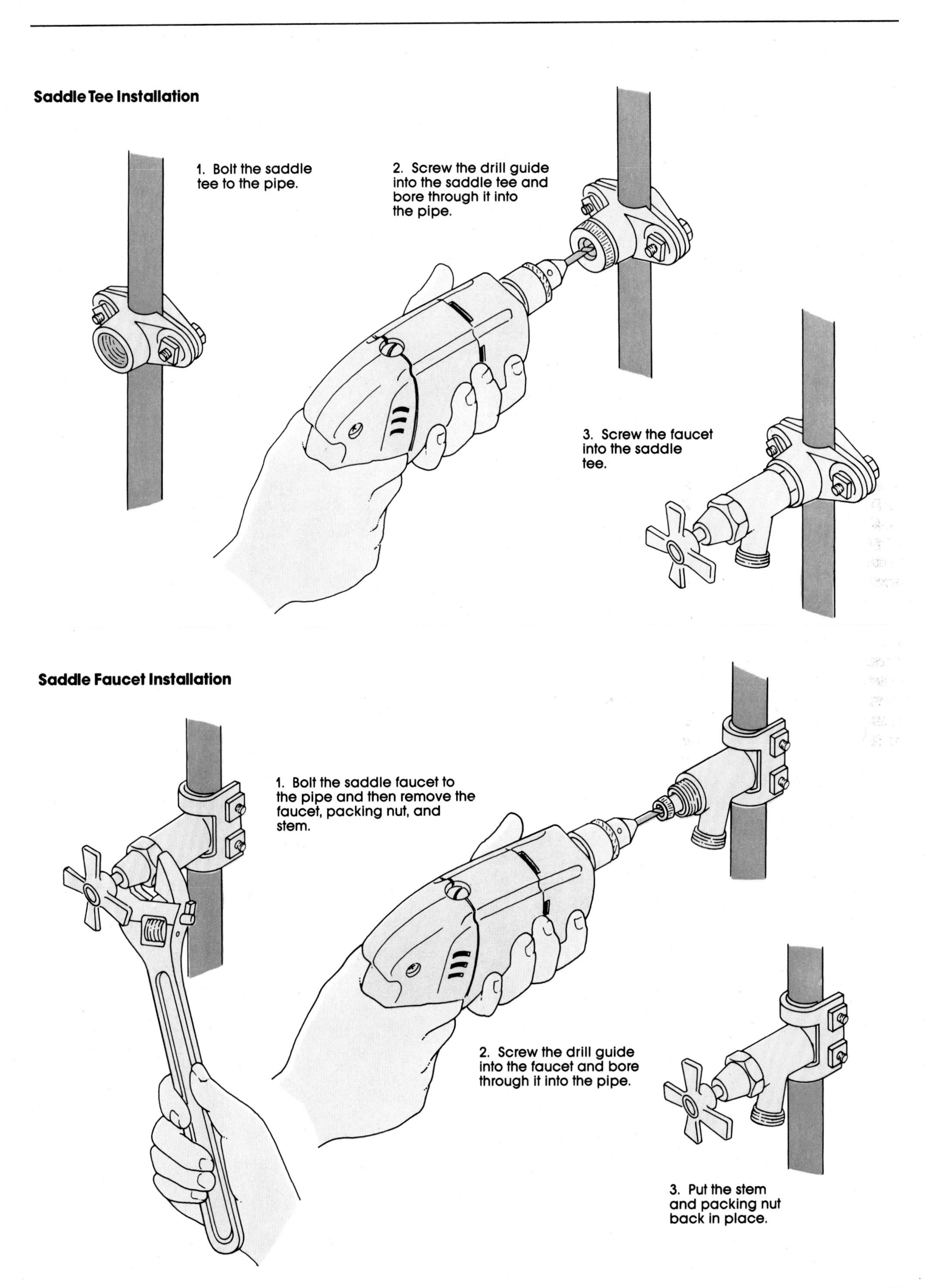

to turn off just the section of the system near where you are working or you may have to turn off the main valve. If your stub is iron pipe, remove the cap with a pipe wrench and clean the threads with a wire brush. Slide the escutcheons onto the stubs, apply pipe compound or teflon tape to the threads, and screw on the valves. When they start to get tight, stop turning so the valve outlet is straight up. Don't turn them too far so that you have to back off. Turning tight squeezes the pipe joint compound out of some threads, and backing off may cause a leak.

If you have a copper tubing system, there will either be a cap soldered onto the stub or the stub will have been crimped flat and soldered. Use a tubing cutter to cut the tubing off about 1½ inches from the wall. Slide

Replacing Bathroom Faucets

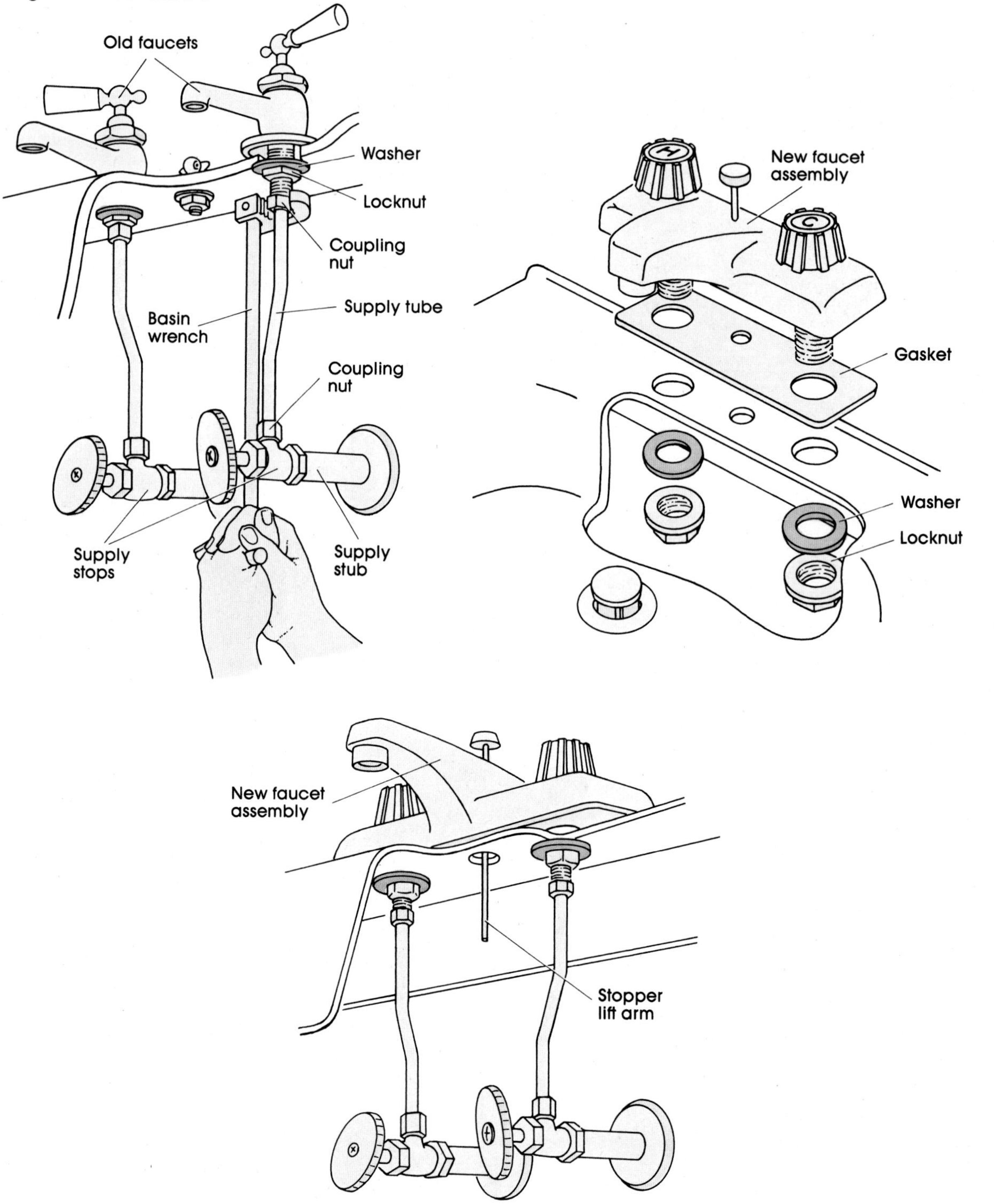

the escutcheons onto the pipes, then the coupling nut, the compression ring, and finally the valve itself. Hold the outlet of the valve up and slide it over the compression ring. Turn the nut onto the threads as tight as you can with your fingers and then tighten it with a wrench. It will usually make a squeaking sound when it seats. When you hear the squeak or it "feels" like it is snug, stop turning so you don't damage the compression ring.

When both stop valves are in place, turn their handles clockwise until they are tight. This closes the valves so the water is turned off right there. You can turn the main valve back on to see if there are any leaks in your new work. Then you can go on to install the faucet assembly and reconnect the supply tubes (see page 54).

Installing a Supply Stop

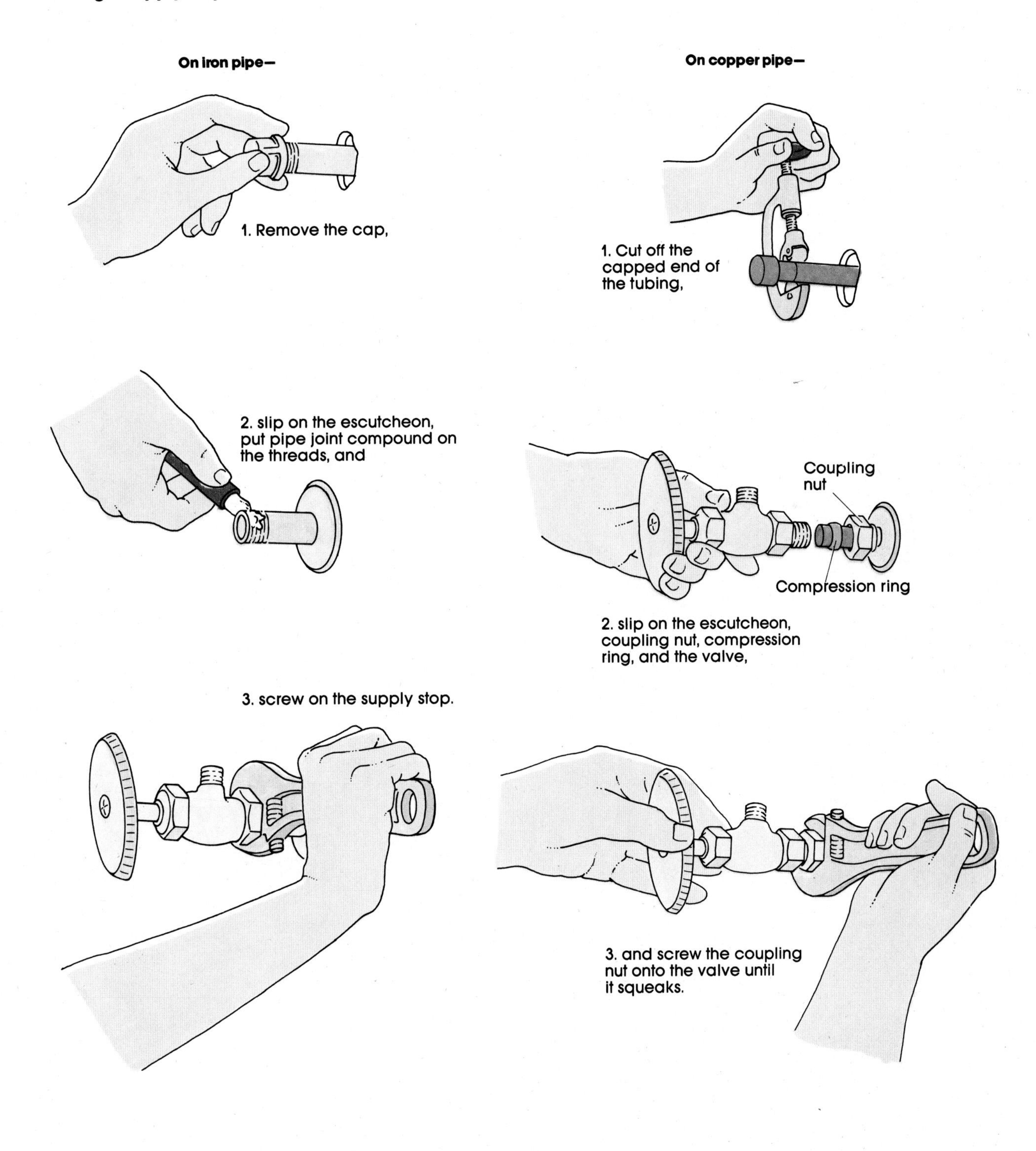

FIXTURES & APPLIANCES

Here is a simple guide to repair and installation of a sink, bathtub, toilet, shower, garbage disposer, dishwasher, or washing machine. Maybe you plan to install and insulate a new water heater. Or perhaps you want to know more about solar water heating systems.

Over the last century, improvements in plumbing fixtures and appliances have made our lives safer and less complicated. Yet many of the designs themselves have become more and more complex.

At least the appliances, such as dishwashers and washing machines, have become more complex. They are a veritable maze of automatic timers, pumps, valves, switches, and pipes. The performance and quality of these devices are not covered by building or plumbing codes, but competition among manufacturers has encouraged steady progress over the years. When selecting appliances, compare as many as possible. Talk to friends and relatives about their experiences with the reliability and durability of various brand names. And check what the current consumer publications have to say about them.

Do not pay a lot of extra money for added complexity if you don't need it—it is a waste in more ways than one. For instance, if you are the kind of person who divides all your dirty clothes into two piles (white and other), you don't need 23 different fabric and temperature settings on the washer. The extra settings not only cost more in the initial investment, but they also require additional parts that can malfunction. There are very few repairs that you can make yourself on these kinds of appliances (see page 74). So it pays you to purchase the simplest machine that will do the job you want done.

Fixtures and Your Health

Plumbing fixtures—toilets, bathtubs, showers, and sinks—have also been improved steadily over the years. Because these are located at the end of the clean water supply system and the beginning of the wastewater drain system and because they influence everyone's health and well-being, building and plumbing code standards control their design and quality. The code requires that plumbing fixtures be made of high-quality materials without defects of any kind. The outer surfaces must be smooth, easy to clean, and nonabsorbent. The inside surfaces must be smooth and free from any ridges, nooks, or crannies that could collect waste material.

Fixtures that meet the code are made of vitreous china, porcelainized or enameled cast iron or stamped steel, stainless steel, and some plastics. The plastics are confined to gray-water fixtures like bathroom sinks, tubs, and showers. Black-water fixtures, like toilets and kitchen sinks, must be made of the other impervious materials. However, a standard for plastic toilet tanks and bowls is being developed, and many plastic toilet tanks are already in use. Fixtures that are constructed on the site out of tile or other materials, are regulated by a separate section of the code and must be approved by an inspection at various stages of their installation.

Plumbing fixtures must always be located in rooms that are adequately ventilated. If there is no window or the window cannot provide the ventilation required by code, a duct and fan must be installed. Usually such a fan is connected to the light switch so the fan is always on when the room is in use.

Codes also require a minimum number of fixtures for each one- or two-bedroom residence. The minimum for any living unit is one toilet, one lavatory (bathroom sink), one bathtub or shower, one kitchen sink, and provision for the installation of a washing machine. Additional bathrooms or half-baths are required for additional bedrooms—the number varies with local codes.

Another part of many fixtures that is regulated by code is the overflow, which prevents flooding over the rim of the fixture. Bathroom sinks and bathtubs are the most common fixtures with overflows, although some of the newer lavatories are being designed without them.

◀ Plumbing fixtures and appliances influence your health and well-being, so choose and maintain them well.

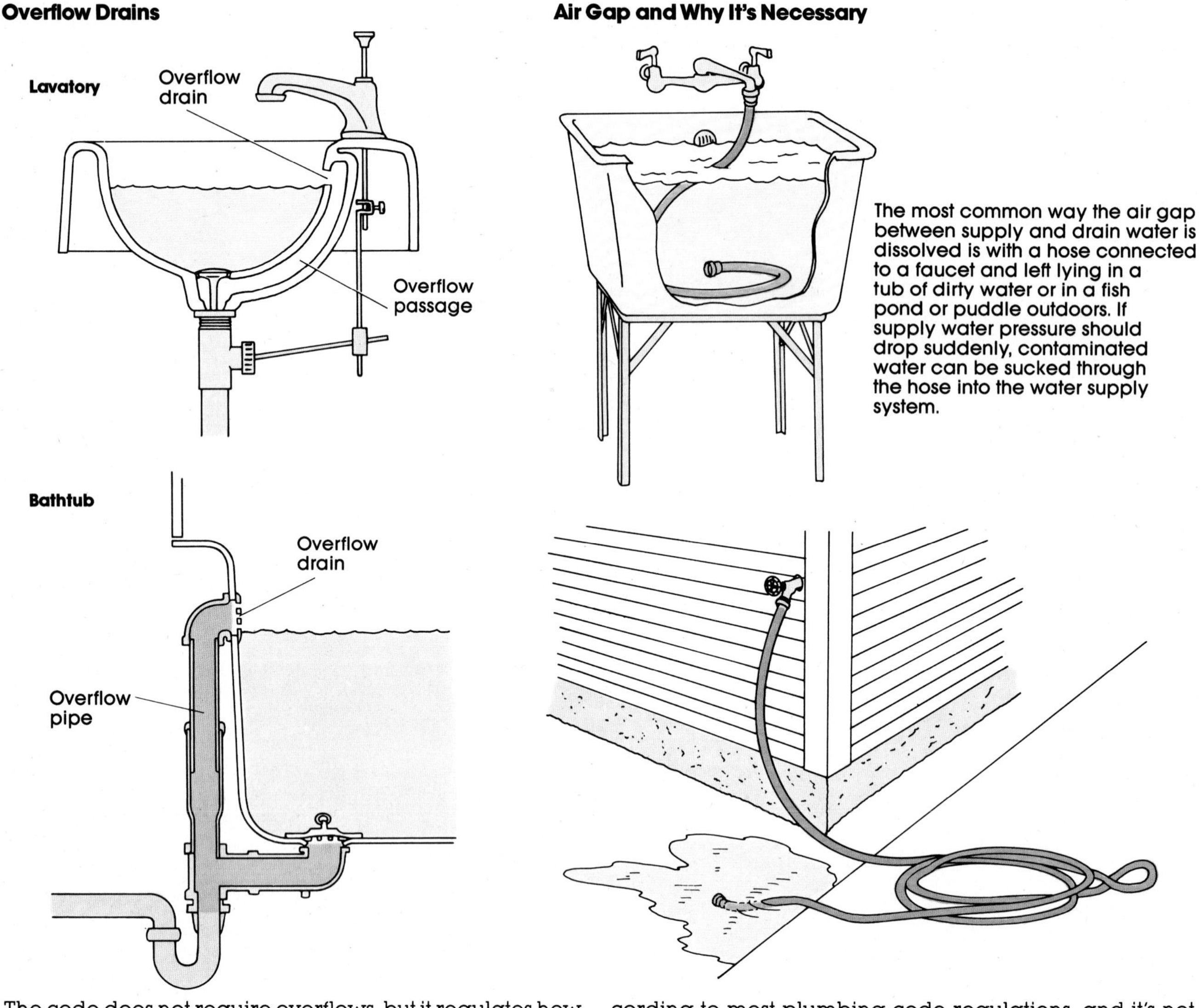

The most common way the air gap between supply and drain water is dissolved is with a hose connected to a faucet and left lying in a tub of dirty water or in a fish pond or puddle outdoors. If supply water pressure should drop suddenly, contaminated water can be sucked through the hose into the water supply system.

The code does not require overflows, but it regulates how they work. An overflow passage, whether it is built into the wall of the fixture or connected to the drain by a pipe, must empty completely. No water may remain in the overflow system at any time. The waste pipe must be designed so water never rises into the overflow. And the overflow passage or pipe must connect to the drain on the fixture side of the trap. Connecting it on the other side of the trap would make the trap useless.

■ A WORD OF WARNING: A cross connection is a configuration of plumbing that connects the potable water supply system to a source of contaminated water. No one, of course, would knowingly make this sort of connection, but they are often made accidentally. The most common cross connections are made with a hose. For example, a garden hose connected to a faucet may be left with its end lying in a fish pond or in a puddle of standing water. An indoor plant-watering hose connected to a sink faucet or the spray-hose attachment on a kitchen faucet assembly or in a bathtub can make a cross connection if it is left lying in a sink or tub of dirty water.

Virtually all sinks, toilets, faucets, and other plumbing fixtures and appliances are now manufactured according to most plumbing code regulations, and it's not likely that you will make a cross connection by installing any of them. However, this wasn't always the case, and it is possible that cross connections exist in older installations. If the flush valve in your toilet is below the water level when the tank is filled or if the nozzle of a faucet is below the rim of a sink or tub, you have a cross connection that should be corrected. Plumbing codes now insist on at least one inch of vertical space between a faucet and the rim of the fixture it serves.

A cross connection doesn't necessarily contaminate the water supply. However, the potential for contamination is there if the pressure in the water supply system should drop below atmospheric pressure. Then the contaminated water would be sucked into the supply system and be in the pipes ready to go into your glass the next time you turn on the water for a drink.

Another source of contamination in a water supply sytem is an underground leak. As long as the pressure in the supply line is higher than the pressure of the groundwater, the leak continues. A drop in pressure, however, could cause the groundwater to be sucked into the system through the hole in the pipe.

TOILETS

A flush toilet is probably one of the most frequently used devices of modern man. People who live in a locality where toilets are the norm probably use them several times each day, from the time they are about a year old. In spite of all of this use, the mechanism remains a mystery to many. If you are one of those people who must call a plumber whenever trouble occurs, read on and the mystery will be solved for you.

When you flush the toilet, a lever lifts a ball or flapper from the valve seat at the bottom of the tank, allowing the water that has been stored in the tank to rush into the bowl. The water rising in the bowl flows over the high point in the trap creating a siphon effect that sucks the contents of the bowl through the trap into the sewer line. As the water level in the tank drops, the float ball drops with it, opening the supply valve (ball-cock) to allow more water into the tank. When the tank is empty, the tank ball or flapper falls back into place in the valve seat so the tank can refill. As the tank is filling, a small stream of water—too small to create a siphon effect—squirts from the bowl-refill tube, down the overflow tube and refills the toilet bowl to the level of the trap. When the tank is filled to the water level line, usually marked by the manufacturer, the float ball closes the ball-cock, and the toilet is ready for its next use.

If you notice your toilet is not flushing completely or that it fills almost to the rim and then drains more slowly than usual, start working on it right away. It has probably developed a partial clog, which it is much easier to clear than a toilet that is completely stopped up. If you get a clog in your toilet, it is probably an emergency (see page 20 in the "Emergency Repairs" section).

Repairing a Toilet

Problems other than clogs are usually associated with the mechanism in the tank. One of the most common problems is water continuing to run in the tank when it should have shut off. First, remove the cover from the tank and put it carefully in a secure place. It can easily be broken. Flush the toilet and watch how the mechanism works. As the tank refills, you will see that the water rises too high, begins flowing into the overflow tube, and doesn't shut off. The water doesn't shut off either because the ball doesn't rise high enough to close the ball-cock valve or the valve is defective.

Pull up on the ball. If the water shuts off, your problem is either that the ball has water in it or it is set too high. Unscrew the ball from the end of its rod and shake it. If it has water in it, replace the ball with a new one. If there's no water in the ball, the rod holding the ball needs to be adjusted lower. Some mechanisms have an adjustment screw on the ball-cock that you can turn to lower the rod. If not, bend the rod to lower the ball. The ball height should be adjusted so the water level is ½ to 1 inch lower than the top of the overflow tube or at the water line marked in the tank by your manufacturer.

If when you pull up on the ball you feel pressure and the water still doesn't shut off, the problem is the valve. If the valve looks like it is in good condition, it may be just a worn washer, which you can change just like the washer

How to Bend a Float Ball Rod

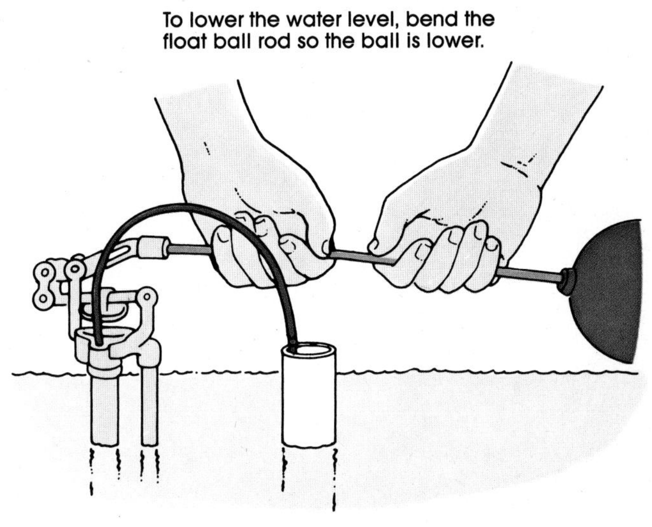

in a faucet (see page 53). If the valve is corroded or appears broken, don't try to fix or replace it with a new one; install a new self-contained plastic flush mechanism (see drawing).

To install this type of mechanism with an adjustable float cup, close the shutoff valve on the water supply line beneath the toilet tank. Flush the toilet to empty the tank. Slip an adjustable wrench or pair of locking pliers onto the nut at the bottom of the ball-cock assembly. With another wrench, loosen the nut under the tank that holds this assembly to the tank. Remove the old assembly and put the new plastic one in its place. Before tightening down the new assembly, be sure the washers are in good shape—not cracked or hardened.

Another common problem with toilets is the tank ball or flapper on the end of the trip wire not seating properly in the drain. The result is that either the tank never fills up or if it does, there is a constant and irritating gurgle. To correct this, flush the toilet to empty the tank and then hold the tank ball up with a piece of string or wire to keep the water shut off. Unscrew the tank ball from its guide rod, clean the valve seat, and screw on a new ball. Or better yet, replace the tank ball with a flapper. The flapper is less prone to misalignment and very easy to install. These flappers are available in any plumbing or hardware store. They just clip around the base of the overflow pipe, and a chain attached to the flapper hooks onto the flushing lever.

Correcting a Sweating Toilet Tank

Moisture may build up on tanks when cold water flowing into the tank cools the porcelain enough to condense water from the warm bathroom air. This is called sweating. It can be more than just a nuisance. The dripping moisture can loosen floor tiles and soak the subfloor, resulting in rot.

The easiest solution, and an effective one unless the water coming into the tank is extremely cold, is to line the inside of the tank with a layer of insulating foam. Half-inch thick styrofoam or foam rubber works quite well. Drain the tank and wipe it completely dry. Then use epoxy

Replacing a Float Ball with a Plastic Mechanism

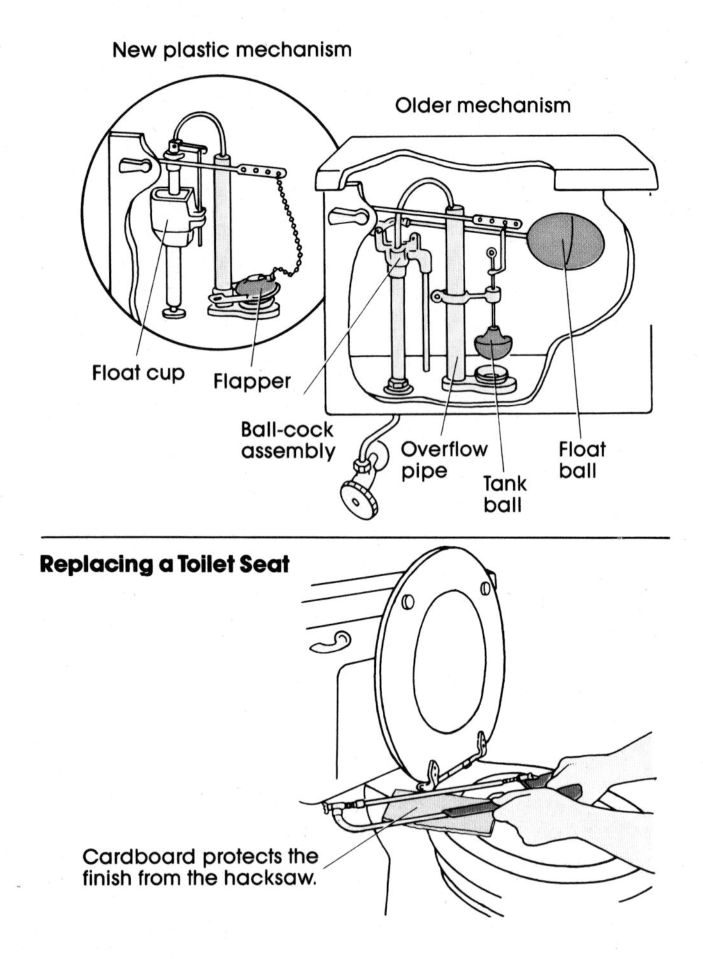

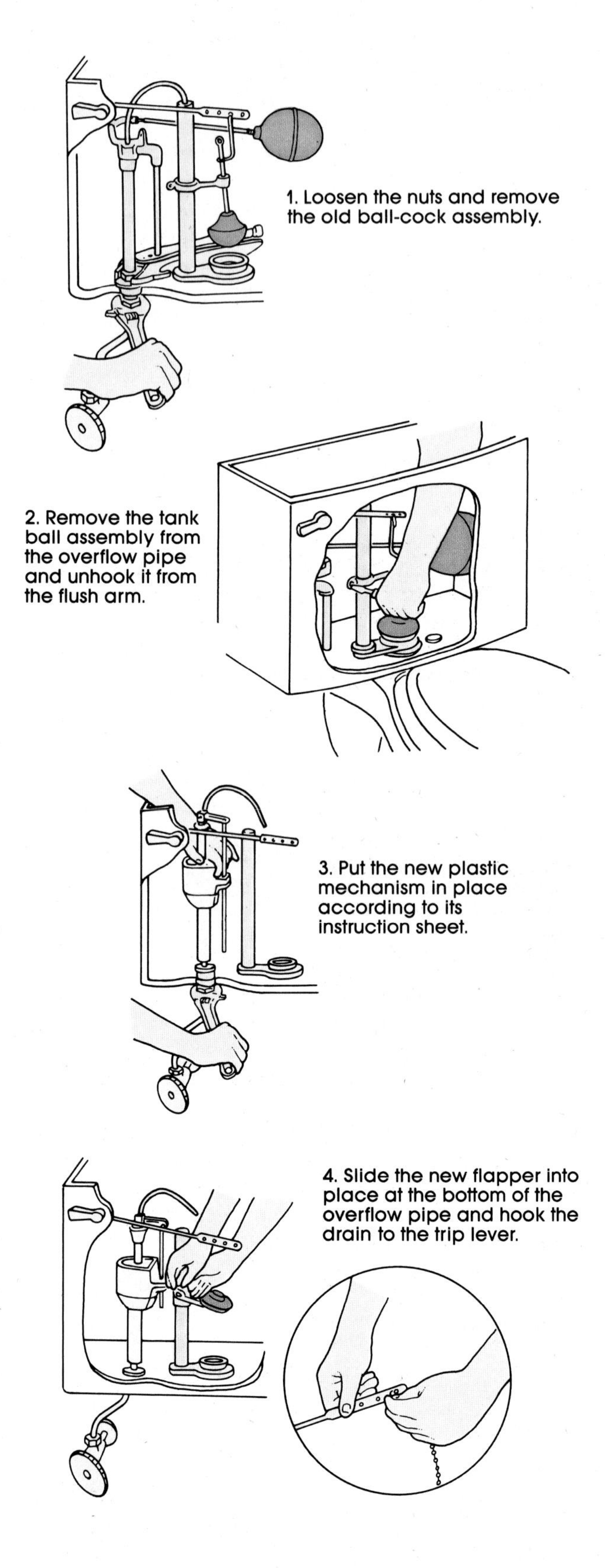

resin cement to affix the insulation to all four sides. It should reach well above the water line and not interfere with any of the mechanisms. Let the cement dry thoroughly before refilling the tank.

If the tank still sweats, you can tap into a nearby hot-water line and let hot water mix with the incoming cold water. Put a reducing tee in both the nearby hot-water line and the cold-water supply line to the toilet. Then connect them with a mixing valve and a pipe one size smaller than the existing lines (see drawing).

Another alternative that would cost less in the long run than heating a lot of extra water would be to buy a replacement tank unit. They come already insulated and complete with all the internal mechanisms for about $50.00. They are easy to install and come with an instruction sheet.

Replacing a Toilet Seat

The only problem here is that occasionally the bolts holding the seat will become corroded and be difficult to remove without damage to the finish on the toilet. If they don't come loose easily, squirt some penetrating oil on them and let them set overnight. If they still won't budge, even when you use a socket wrench, saw them off with a hacksaw. Put a piece of cardboard between the saw and the porcelain so it isn't damaged.

When you select a new toilet seat, it is a good idea to have the dimensions of the bowl and the position of the bolt holes. Either carry the old seat with you to the store or trace the outline of the bowl and the position of the bolt holes on a piece of heavy cardboard and take that with you.

Replacing a Toilet

If you have an old-style toilet that is noisy or if your old toilet has been giving you trouble on more than rare occasions, you might consider replacing it with a new-style toilet. Replacement is really not all that difficult, and it is certainly less expensive than calling the plumber all the time.

Each manufacturer's product has its own differences, but toilets are all variations on four basic designs. As you might expect, the newer, quieter, more efficient designs are also more expensive. A toilet lasts a long time, however, so you may feel the more expensive style is an economy in the long run.

The oldest style toilet still in use is the *wash-down* style. It is noisy, not very efficient, and has a minimum amount of water surface in the bowl so the porcelain area of the bowl is easily soiled. In many areas, wash-down toilets are no longer allowed by code, and you probably couldn't find one even if you wanted one.

The *reverse trap* toilet is a step up in efficiency and quietness, it has a deeper trap, and more of the bowl area is covered with water. These may be available in some places as "economy" toilets, but even they are now becoming harder to find.

The *siphon jet* toilet was a great leap in efficiency. It is identifiable by a small hole below the water line used to direct a jet of water into the trap, starting a quick, powerful siphon that pulls the water and waste material from the bowl. This toilet is quieter and uses less water than the reverse trap toilet.

By far the quietest, most efficient, and most expensive toilet made today is the *siphon-action* or *one-piece* toilet. Instead of a little hole and a jet of water, this toilet has an elongated hole at the side of the bowl. Water from this opening starts the siphon with a swirling movement. This style has steeper sides and a rounder bowl bottom, providing maximum water surface area and making it very difficult to soil the sides of the bowl.

For information on various alternative toilets, such as waterless and up-flushing toilets, see page 17.

Removing the old toilet. Replacing a toilet is more time-consuming than it is complicated. To have plenty of time, allow yourself four or five hours for the job. Start by turning the water off under the tank and then flushing the toilet. Use a sponge to remove as much water as you can from the bowl and the tank. Disconnect the water supply line from the bottom of the tank. If the water supply line goes to the tank directly without a supply stop valve, plan on installing a new supply stop and connecting it to the new toilet with a flexible connector.

If your old tank is wall-mounted, remove the spud pipe or elbow connecting it to the bowl. Unscrew the nuts from the hanger bolts inside the tank and lift the tank out

Kinds of Toilets

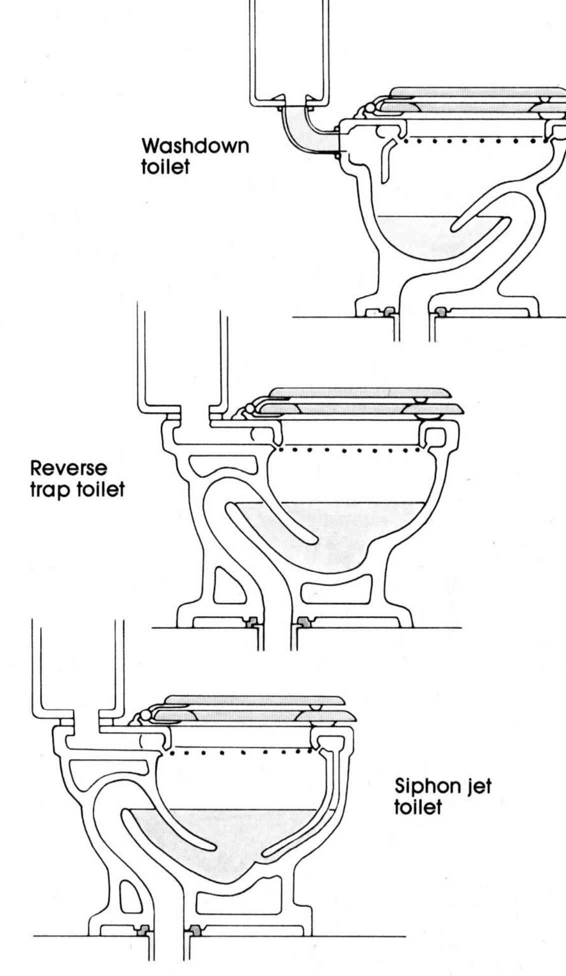

of the way. If the tank is mounted on the bowl, unscrew the bolts that go through the bottom of the tank and through the back lip on the bowl, and you will be able to lift it right off.

Before removing the bowl from the floor, measure the rough-in distance from the wall to the two bolts at the base of the bowl. If you have an old toilet and there are four bolts, measure the distance to the back ones. It is normally 12 inches. If your measurement is different, explain this to the person selling you a toilet, who will recommend a proper replacement.

Unscrew or pry the caps from the bolts and remove the nuts. Then straddle the bowl and rock it gently from side to side to loosen the wax seal. When it's free, lift the bowl straight up and hold it that way, so water doesn't spill from the trap as you carry it carefully out of the house.

With a knife or other convenient tool, remove all the wax and other accumulated stuff from the floor flange. Do this job thoroughly to avoid leaking around the flange when you install a new wax gasket.

Installing the new toilet. Take your new toilet out of the box and lay the bowl upside down on a bed of newspapers. The vitreous china will scratch, so do your best

Removing an Old Toilet

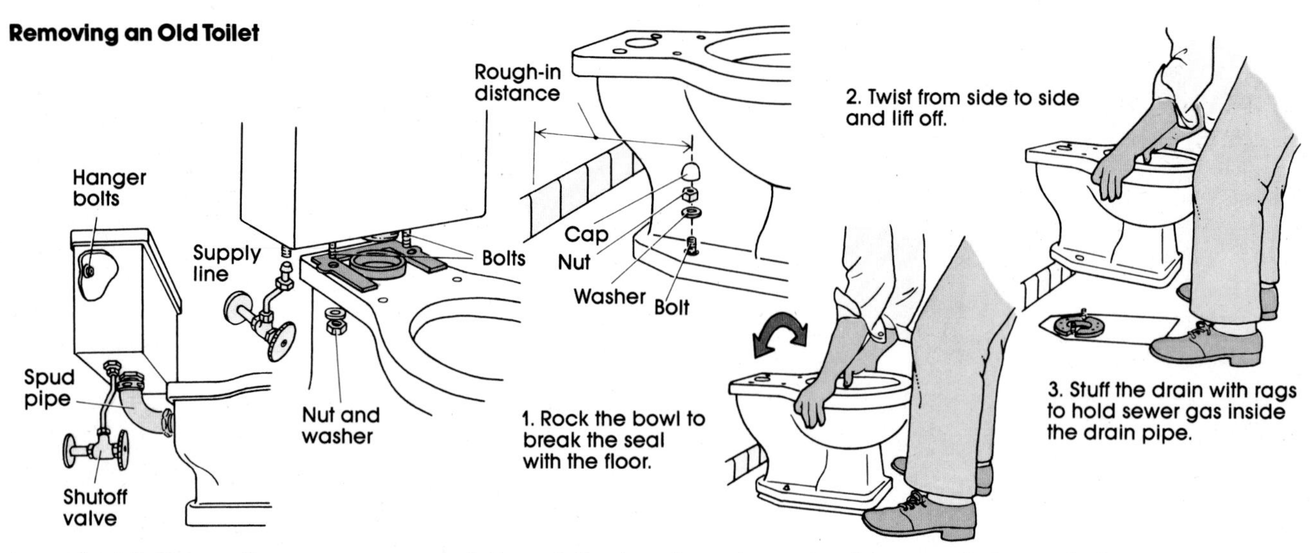

to protect it. Put on the new wax gasket. Press it firmly and evenly over the horn (water outlet), with the tapered side fitting against the base of the toilet. If the floor flange is recessed into the floor, get a wax gasket with a plastic sleeve attached that fits down into the flange. Slip new hold-down bolts into position on the flange and, if necessary, use plumber's putty to hold them in an upright position. Put a thick layer of plumber's putty all around the edge of the toilet base.

After double checking to be sure there is no more packing paper inside the bowl, turn the toilet over and set it smoothly and evenly in place on the flange. Twist it back and forth slightly to smooth out the putty and then sit on it to force the gasket down onto the flange. Set a carpenter's level on the bowl and make sure it is level both from front to back and from side to side. Slip thin metal shims under the base to get it level and to eliminate any rocking. When it's firmly in place and level, tighten the nuts on the hold-down bolts. Be very careful here. You want the nuts snug, but not too tight. If you do tighten them too much, they can crack the bowl, and there is no plumbing supply house in the world that will give you a refund or a new bowl if you crack it by tightening the nuts. You will have to buy a new toilet. Double check the level to be sure it hasn't moved.

To mount the tank on the bowl, place the rubber cushion on the bowl so it lines up with the bolt holes. Push the cone-shaped rubber gasket over the tank's flush outlet, and set the tank in place. Put the rubber washers on the bolts and slip them through the bottom of the tank and the bowl lip. Put rubber washers on the ends of the bolts and screw on the nuts. Again, tighten them until they are snug but not too tight.

All that remains now is to hook up the water supply line. If the new tank is lower—which is often the case when the old toilet had a wall-hung tank—put an elbow on the *stub-out*, a 4- to 6-inch nipple on the elbow, and then another elbow and close nipple to which you'll attach the supply stop (see the drawing). Connect the supply stop to the tank with a flexible connector. If the stub-out is copper tubing, you will have to sweat on a brass adapter or use a compression fitting.

Installing a New Toilet

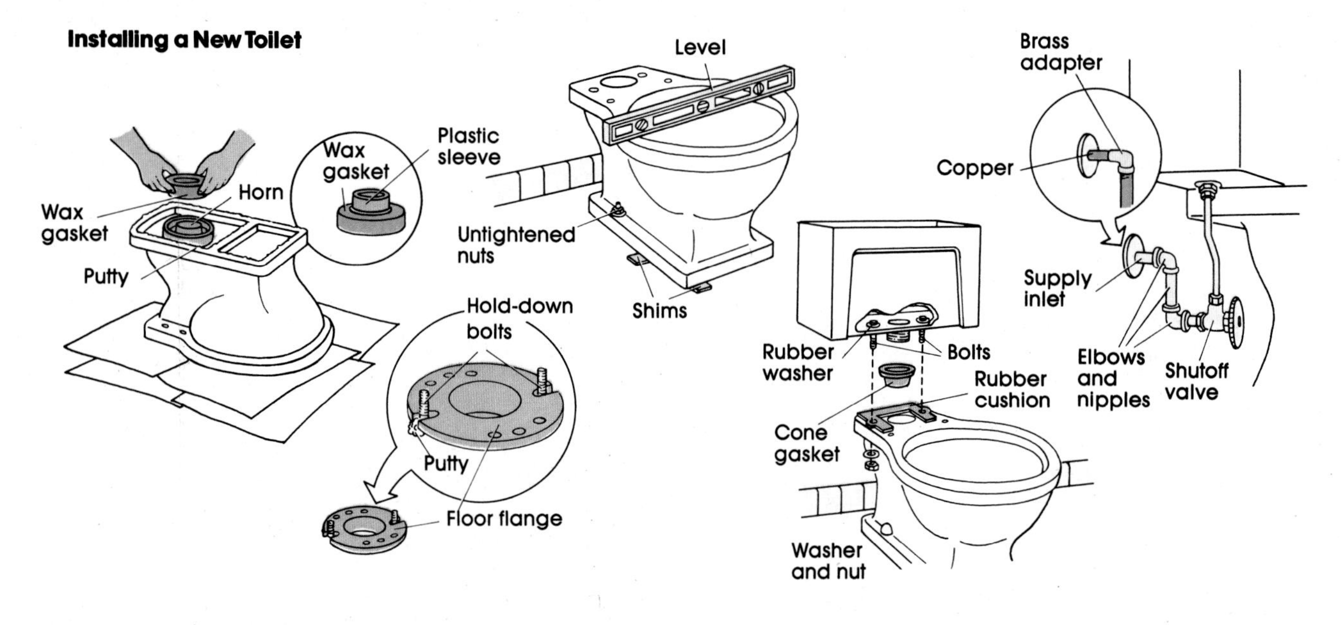

LAVATORIES

Repairing a Bathroom Sink

This is a no-win proposition. If the sink itself is chipped or cracked or part of it is broken off, it can be repaired, but the result is not usually very satisfactory. Consider any repair to be temporary—designed to last only until you get around to putting in a new sink.

A small chipped-off area or a crack can be filled or covered with appliance repair enamel or plastic tub and tile sealer. Be sure the area to be fixed is very clean, dust-free, and dry before applying anything. The enamel comes in a small tube with a brush attached to the cap. It comes in white and colors to match enameled kitchen and laundry appliances. These colors will sometimes match—or come close enough to—the colors of bathroom fixtures. Get the little brush full of enamel and let it flow into the crack or chipped place. The enamel will have a rounded edge that will be visible when it dries, but we told you it wouldn't be perfect. The plastic sealer comes in a tube and is normally available in white only. It is designed to repair the grout around bathtubs and showers. Squeeze a little into the crack or chip and smooth it with your finger. It, too, will show when you've finished, but you can hope it won't be as apparent as the chip or crack. And even if it is seen, it will look like you have tried.

If a piece is broken off, like a corner or an edge, or you have a chip that fits in the place it came from, it can be replaced with epoxy resin or super glue. Again, be sure the surfaces are very clean and dry. Coat one surface or both, depending on the label directions, and press the pieces together firmly. Hold the repair together with masking tape for an hour or so (even if the directions say the glue sets instantly) and don't get water on it for several hours. Overnight is best. This method also works for broken ceramic soap dishes, toothbrush holders, and toilet paper holders.

Removing an Old Lavatory

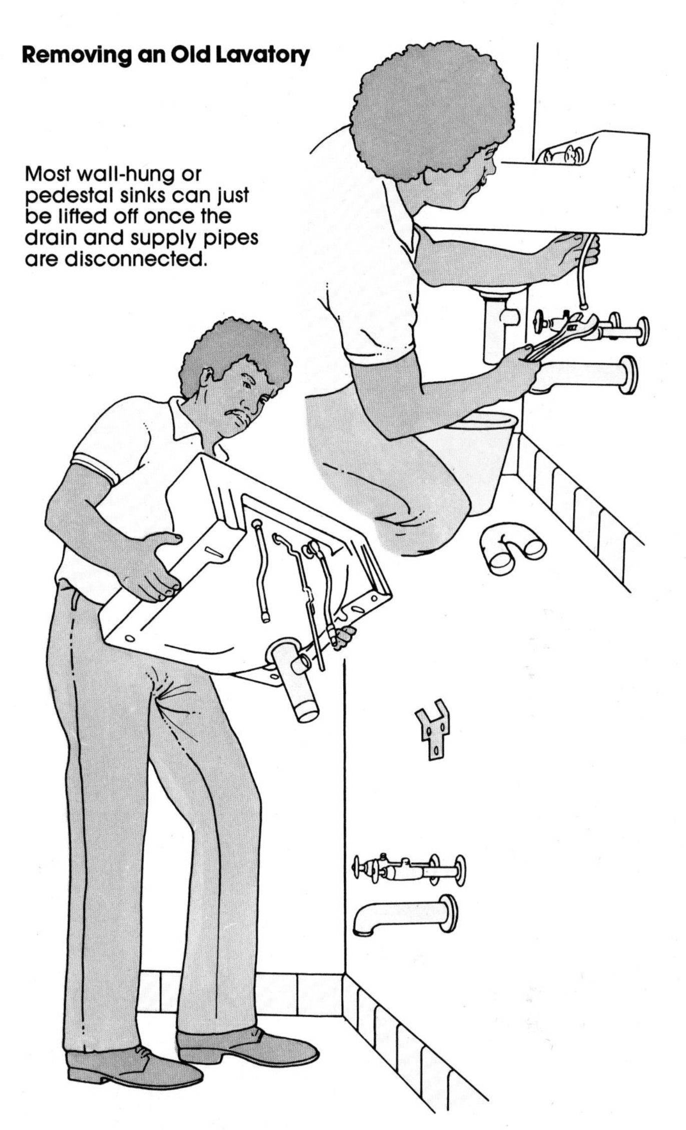

Most wall-hung or pedestal sinks can just be lifted off once the drain and supply pipes are disconnected.

Removing an Old Lavatory

The first step is to disconnect and remove the drain pipe and supply pipes. Remove the P-trap by loosening the compression nuts and sliding the pipes apart. If the trap has a plug at the bottom, remove it and let it drain into a pail before you take it apart. If it doesn't have a plug, set a pail beneath it to catch the water as you take it apart. Close the supply stops and remove the supply tubes by unscrewing the fittings at the stops and at the bottom of the faucets. If you plan on using those same pipes on the replacement sink, protect the finish by wrapping tape around the nuts before you apply the wrench or pliers.

Old-fashioned pedestal or wall-hung sinks are not usually screwed down. They are heavy enough that their weight keeps them firmly in place and, of course, the pipes hold them, too. With all the pipes disconnected, you should be able to just lift the sink off the pedestal or wall bracket.

A few words of caution are important here. Old ceramic or china fixtures are very heavy. It would probably be in your best interest to have a helper assist with the lifting and carrying. If you are removing a very old wall-hung sink, the wood behind it, to which the metal hanger is screwed, may be rotten. The pipes may be holding the sink up. Have your helper hold onto the sink when you remove the pipes so if that is the case, the sink won't fall on you.

The pedestal may be just standing on the floor, it may be bolted, or it may be grouted into the tile. If it's set into the tile, you may have a section of flooring to replace as well as a fixture.

Installing a New Lavatory

There are still wall-hung lavatories—you may even be able to find one that will hang onto your old wall bracket. But as long as you are going to the trouble of replacing a fixture, you should select one that gives your bathroom the look you want. You can probably find one that is more utilitarian as well. The four basic kinds of lavatories are wall-hung, leg-stand, pedestal, and cabinet-mounted. The leg-stand type is hung on the wall but has adjustable legs in the front for additional support and for decoration. The cabinet-mounted sink comes in a large variety of shapes and sizes. You can buy one already installed in a cabinet or ready to install in a cabinet you buy separately or make yourself.

Installing a Wall-Hung Sink

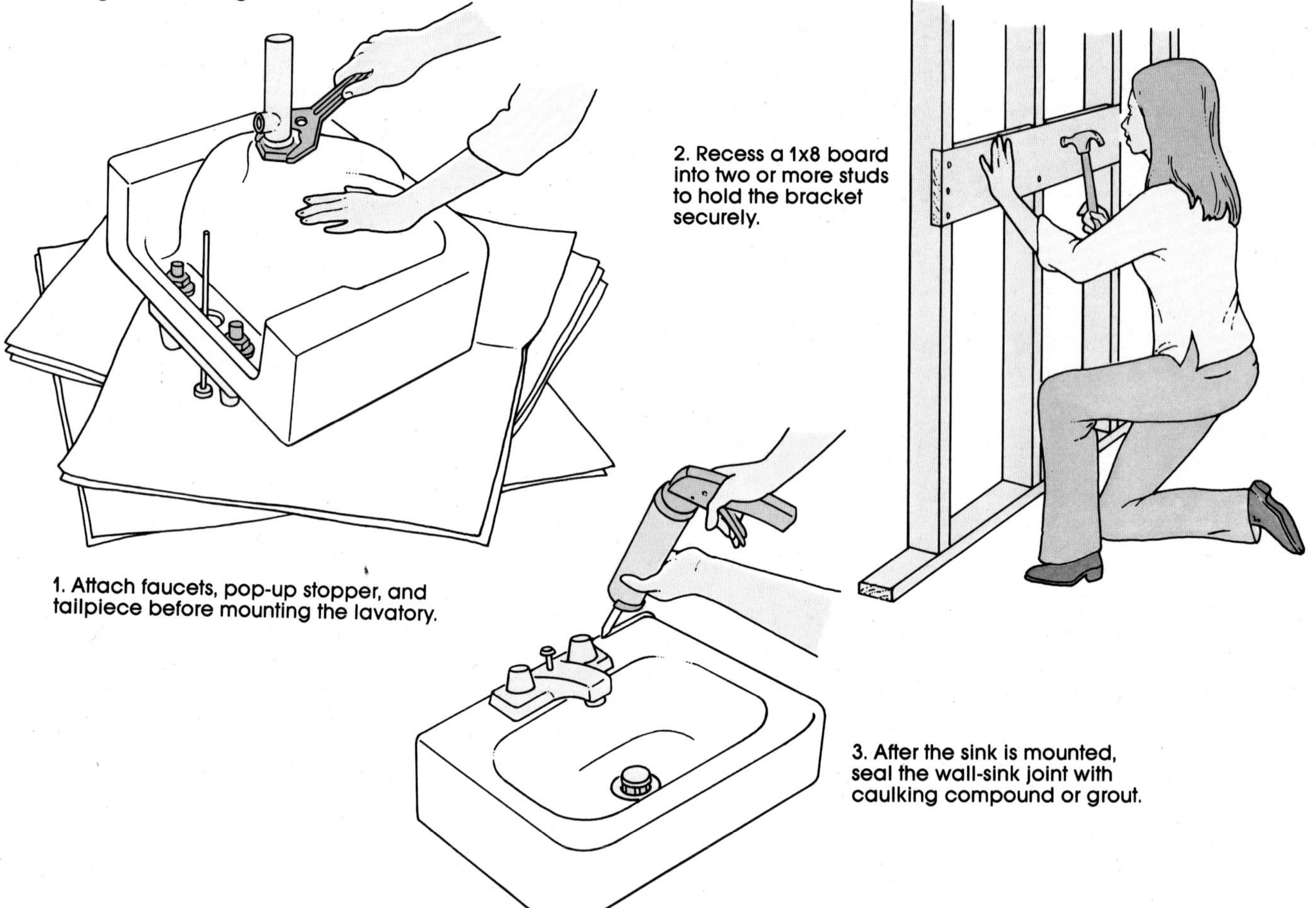

1. Attach faucets, pop-up stopper, and tailpiece before mounting the lavatory.

2. Recess a 1x8 board into two or more studs to hold the bracket securely.

3. After the sink is mounted, seal the wall-sink joint with caulking compound or grout.

Each sink will have an instruction sheet with diagrams and templates for the placement of any holes, supports, or brackets that are needed. The faucet assembly, pop-up stopper, and tailpiece for the drain are easier to attach before the new lavatory is mounted. Set it on the floor with several layers of newspaper or cardboard under it to protect it from scratches. Attach the faucet assembly as shown on page 54 and screw the pop-up drain stopper and tailpiece in place on the bottom.

The wall-hung sink is very easy to install. You simply hang it on a bracket. The bracket, however, must be attached with great care to provide a solid support. It not only must hold up the sink but also the weight of the water it holds and your weight against it as you lean forward to look in the mirror for another wrinkle.

To hold the bracket adequately, nail a 1x8 board directly across two or more studs. Recess it into the studs so it will be exactly flush with the surface of the wall after the sheetrock or plaster is applied. The top of a standard sink or the counter top that holds one is 31 inches. They can be installed as high as 35 inches for a taller-than-average family or one that has no children. Look at the diagram enclosed with the sink to see how high from the floor the 1x8 board must be to hold the bracket and the sink at the height you want it.

The place where lavatories are attached to the wall must be sealed with caulking compound or grout. Water seeping between the sink and the wall can cause a lot of damage in a very short time.

The leg-stand sink is installed the same way as the wall-hung sink except that the wall bracket need not be as strong. The legs are able to support a great deal of the weight.

Some pedestal sinks just stand on the floor. Their own weight and the pipes hold them in place. Others are bolted to a bracket or flange that has been screwed to the floor. Still others are supported by a wall bracket as well as the pedestal, similar to the leg-stand sink. Place the pedestal, and wall bracket if there is one, as instructed on the sheet accompanying the sink. It will tell you how far the sink must be from the wall and the minimum distance to a wall on the side.

There are many types of cabinet-mounted countertop lavatories and clamp assemblies to hold them in place. There are self-rimming sinks that are set on top of the counter, sinks that have a stainless steel rim connecting them to the counter, under-the-counter sinks, and counters and sinks that are molded all in one piece. Each manufacturer has its own styles and designs, and there are many variations in sizes, shapes, and methods of installation. Each lavatory will have an instruction sheet with any necessary diagrams or templates to help you position and install it. If mounting clamps are needed to hold the unit in place, these, too, will have instructions attached.

In all cases, except when the counter top and sink are one molded unit, you will need to apply a bead of sealing compound or grout where the lavatory and counter top meet. Apply the sealing compound, usually to the rim of the sink, before it is set in place. Wipe off the excess after the unit is secured. Apply grout between a self-rimming sink and the tile counter top after the sink is in place and secured.

The next step is to attach the faucets to the new sink, if you haven't done it already, and hook them up to the hot- and cold-water supply stops.

Next, install the P-trap assembly. The parts you'll need for this are as follows:

Tailpiece. If you've bought a faucet assembly that includes a pop-up drain stopper, a tailpiece with the stopper built in should be in the box with the faucet. Otherwise you will need to buy a tailpiece along with the P-trap.

P-trap. This consists of the trap itself and the waste arm. When these pieces of tubing are put together, they form the "P" shape that gives the fitting its name. Along with the tubing, two machine-threaded slip nuts, one pipe-threaded slip nut, and three rubber washers come with the unit.

Adapter. Since the waste arm and all these parts are 1½-inch tubing, the pipe-threaded slip nut needs 1½-inch pipe threads on the drain stub to make a proper connection. If the stub coming out of the wall is 1½-inch threaded pipe, you won't need an adapter. If the stub is copper tubing or plastic pipe, you will. The plastic pipe adapter is cemented to the stub. The copper tubing adapter is sweated on.

If you didn't attach the tailpiece to the sink before it was set in place and if an adapter must be cemented or sweated to the stub, do these things next. Then, to install the P-trap you first put a slip nut and washer on the tailpiece, slip the longest side of the trap onto the tailpiece, and finger tighten the slip nut onto the trap threads. Second, put the pipe-threaded slip nut and washer on the waste arm, slide the straight end into the drain stub with the bent end pointing down, and finger tighten the nut onto the stub. Put the third slip nut and washer onto the bent end of the waste arm. The trap can slide up and down a little and turn on the tailpiece, and the waste arm can slide in and out of the drain stub. This should give you enough play to manipulate the trap and waste arm until their ends meet and you can screw the third slip nut onto the short end of the trap. Tighten all the slip nuts snugly, but not too tight, with a wrench.

And finally, connect the pop-up stopper assembly if one comes with the faucets you've chosen. Put the stopper assembly together as shown in the drawing and adjust it so it opens and closes properly by loosening the clevis screw and sliding the clevis up or down on the lift rod.

For instructions on installing a new lavatory either next to or in a room adjacent to an existing lavatory, see page 92.

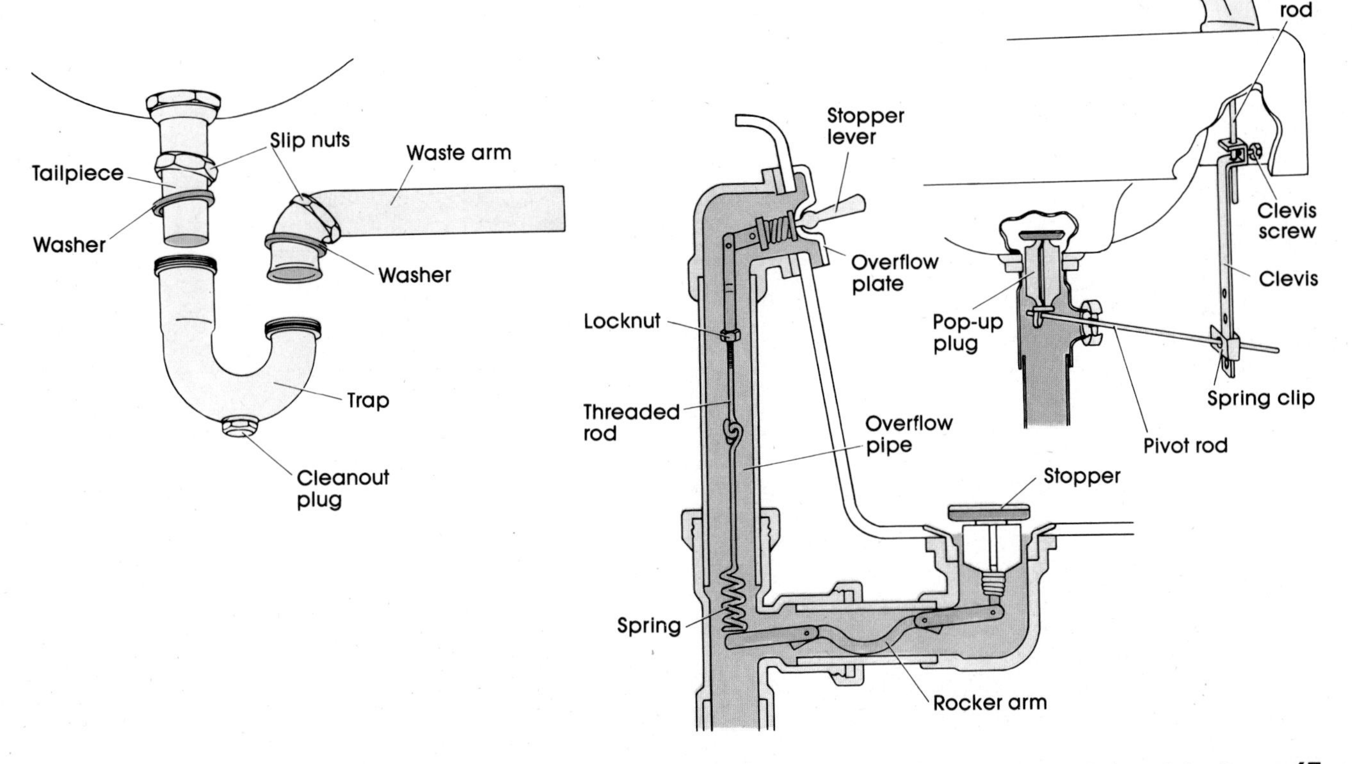

BATHTUBS

Old bathtubs were made of porcelainized cast iron. Later they were made of enameled stamped steel. Now, in addition to cast iron and steel, they are being made of acrylic and fiberglass reinforced plastics. Not only are the new plastics warmer and more comfortable to the touch, they are less expensive, lighter in weight, and easier to fabricate in more intricate shapes.

Repairing a Tub

Bathtubs themselves don't often need repairing. A clogged or maladjusted drain, a chip in the porcelain, or loose caulking around the tub is usually the extent of it.

If the tub drains slowly or not at all, the problem could be the stopper mechanism. In addition to the good old rubber plug, there are two main types of stoppers on bathtubs. One is called a pop-up and the other a trip lever. Both are controlled by a locknut on a threaded rod. Repeated opening and closing of the drain can cause the nut to loosen and move. This can result in a leak in the stopper or in slow drainage. These mechanisms can also collect hair and lint, which slows down the draining. Take them apart and clean them before you do any adjusting. If you have a rubber stopper, the drain is probably clogged with hair. Use a trap-and-drain auger to clear it (see page 20).

Many codes do not allow you to install new trip-lever drains. In this type of drainage arrangement, the plug is in the overflow pipe and works by dropping straight into a seat at the base of the overflow. If the adjusting nut has worked its way up, the plug will not fit tightly in the seat and water will leak through when the plug is supposed to be closed. If the nut has worked down on the threaded rod, the plug would fall into the seat, possibly at some crazy angle, and the only way to get it out would be to fish it out with a piece of bent wire or to take the drains apart. To adjust the nut, unscrew the overflow plate that has the stopper lever on it. Lift the plate and pull the whole mechanism out of the overflow pipe. It usually won't take much of an adjustment to correct the problem. It must be done simply by trial and error.

A pop-up drain stopper has similar linkage to that of a trip-lever drain. Instead of a stopper at the end of the linkage in the overflow pipe, there is a spring. This spring hits the end of a rocker arm that is connected to the stopper in the bathtub drain. The spring pushing down on the rocker arm opens the drain stopper, and the stopper falls into the drain when the pressure of the spring is removed. If the drainage of the tub is slow, do not try to correct it by removing the stopper plug. The plug is designed to keep small objects from being swept into the drain where they can cause clogs. To adjust the drain, unscrew the overflow plate that holds the stopper lever and pull out the linkage. Then pull out the stopper and the rocker so you can check and clean it. Remove any hair that has accumulated on the linkage and check the rubber washer around the edge of the stopper. If the rubber isn't soft and pliable, replace it. When you put the rocker arm back into the drain, be sure the curve of the rocker is down. To make the stopper rise higher in the drain, lengthen the rod by loosening the locknut and unscrewing it a little. To make the plug fit tighter, shorten the rod by screwing it in farther.

If the porcelain on your bathtub is chipped, buy some appliance repair enamel at the hardware store. The enamel and how to use it is described on page 65 for the repair of lavatories.

If the grout around the tub is loose or missing, you'll need a tube of plastic tub and tile sealer and caulking. Clean away all the loose and discolored caulking with a screwdriver or putty knife. Be careful not to scratch the tub or wall tile. Let the crack dry for several hours or overnight or dry it thoroughly with a hair dryer before you apply the plastic caulk. While you are at it, check to see that the grout between any nearby tile is not loose. If you find loose grout, scrape it out, too. Cut the tip of the nozzle for a moderate-sized bead; about 3/16 inch is right. Hold the tube at a 45-degree angle and push it along the edge of the tub in one continuous line as you squeeze. If you can do each side of the tub in a continuous motion without stopping, the line of caulk will usually be much smoother and neater.

Replacing a Tub

Before you decide to replace your old tub, read this and then study your old tub so you'll know what you are getting into. There are free-standing tubs and tubs that are built into two or more walls, usually three. Most tubs are placed on the subfloor while the house is being built and flooring—tile, linoleum, and so forth—is laid so it contacts or is attached to the tub. A free-standing tub (one that's not attached to the walls) cannot be removed without doing considerable damage to the flooring that it contacts. A free-standing old tub that has feet sits on top of the floor, and when you disconnect all the pipes, you can just carry it away—if you are strong enough. If you have a built-in tub, both wall and floor damage will occur when you remove it.

Even if you can remove your tub without damage to the walls and floor, the installation of a new tub will still require considerable planning and rebuilding of the walls and floor where the tub will go. When selecting a new tub, take a very detailed plan of your bathroom to a plumbing fixture showroom. Be especially careful with the dimensions of the area where the tub will go so you select one that will fit with the least amount of rebuilding of walls.

Some of the largest manufacturers make cast iron tubs in the most modern designs and colors. They are the most expensive, the heaviest, and by far the most durable. One will probably last you and your grandchildren until everyone is tired of it. Stamped steel tubs are less expensive, lighter, and still very durable. However, they will chip and dent more readily than cast iron tubs. Plastic or fiberglass tubs are the least expensive, lightest in weight, easiest to install and, of course, the least durable. Although they have a strong resistance to chipping, they scratch rather easily and their glossy surface will dull with time and cleaning. Gritty cleansers will dull them instantly, and once the surface is dulled or scratched it is difficult to ever get it looking good again.

Removing a Bathtub

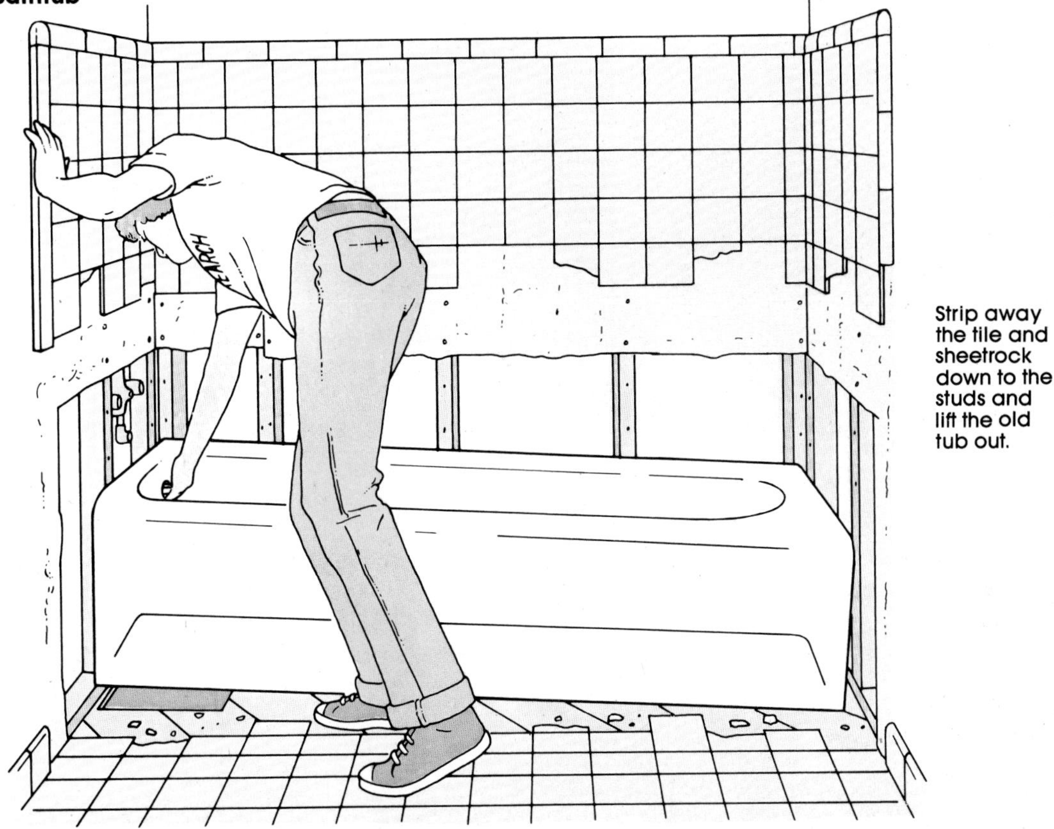

Strip away the tile and sheetrock down to the studs and lift the old tub out.

Removing the old tub. First disconnect all the pipes. If the tub has feet you can just disconnect the pipes and carry it away—with a little help, of course. If it is attached to the floor and wall, you will have to strip away the tile, sheetrock, and whatever down to the studs within several inches of where the tub makes contact. If you have a cast iron tub, you can break it up with a sledgehammer. Be sure to wear eye protection, gloves, and heavy clothing to protect yourself from flying cast iron. A steel tub will have to be lifted out and carried away in one piece. The only way to tell the difference when the tub is in place is to tap on it. The steel will ring or sound hollow, while the cast iron will sound solid like a rock.

Installing the new tub. When the old tub is gone, all the debris is cleared away, and the studs are clean and free of nails, you can proceed with the installation. The new tub should fit exactly between the studs. If the space is too long, build out the wall at the foot of the tub with new studs. This can become a convenient shelf for your sponge and shampoo.

If you had a really old-style tub with the faucets and spout attached to it or if you want the faucets and spout in a different place, you have some plumbing to do, too. All new tubs have only two holes: the drain and the overflow. The faucets, spout, and shower are always mounted in the wall above the tub. All the piping for these must be in place and connected before you put in the new tub.

Select the kind of faucets, spout, and shower head you want and install them according to the instructions accompanying them and the drawings of typical installations shown here. Included will be the hot- and cold-water supply pipes; hot and cold valve assembly; a diverter valve to turn on the shower, either between the faucets or on the spout; the shower pipe with a special elbow and brass screws securing it to a cross brace; and pipe stubs onto which the tub spout and shower head will attach.

Often local building codes will require that there be access to the supply and drain pipes servicing a bathtub. When your bathtub is on the ground floor, this access can be from the basement or crawl space beneath the house. Cut a hole one foot or so by 6 to 8 inches through the subfloor next to the wall under the head of the tub. If your bathroom is on an upper floor (or if you prefer), build an access door or removable panel in the wall of the next room or hall next to the head of the tub (see the drawing).

The tub you buy will have an instruction sheet and any diagrams or templates necessary to place the holes, cleats, and so forth. Most steel and plastic tubs have flanges that are designed to rest on 1 × 4 or 1 × 6 cleats nailed to the studs. Be sure these boards are placed exactly at the right height to hold the flanges and that they are perfectly level. Cast iron tubs won't need these cleats.

Before setting the tub in place, make certain you have plenty of room to work on the drain. Even on an upper floor, you can remove quite a bit of subfloor next to the wall at the head of the tub. It doesn't support anything anyway. Lower the tub carefully into place so the

Tub/Shower Rough Plumbing for Water Supply

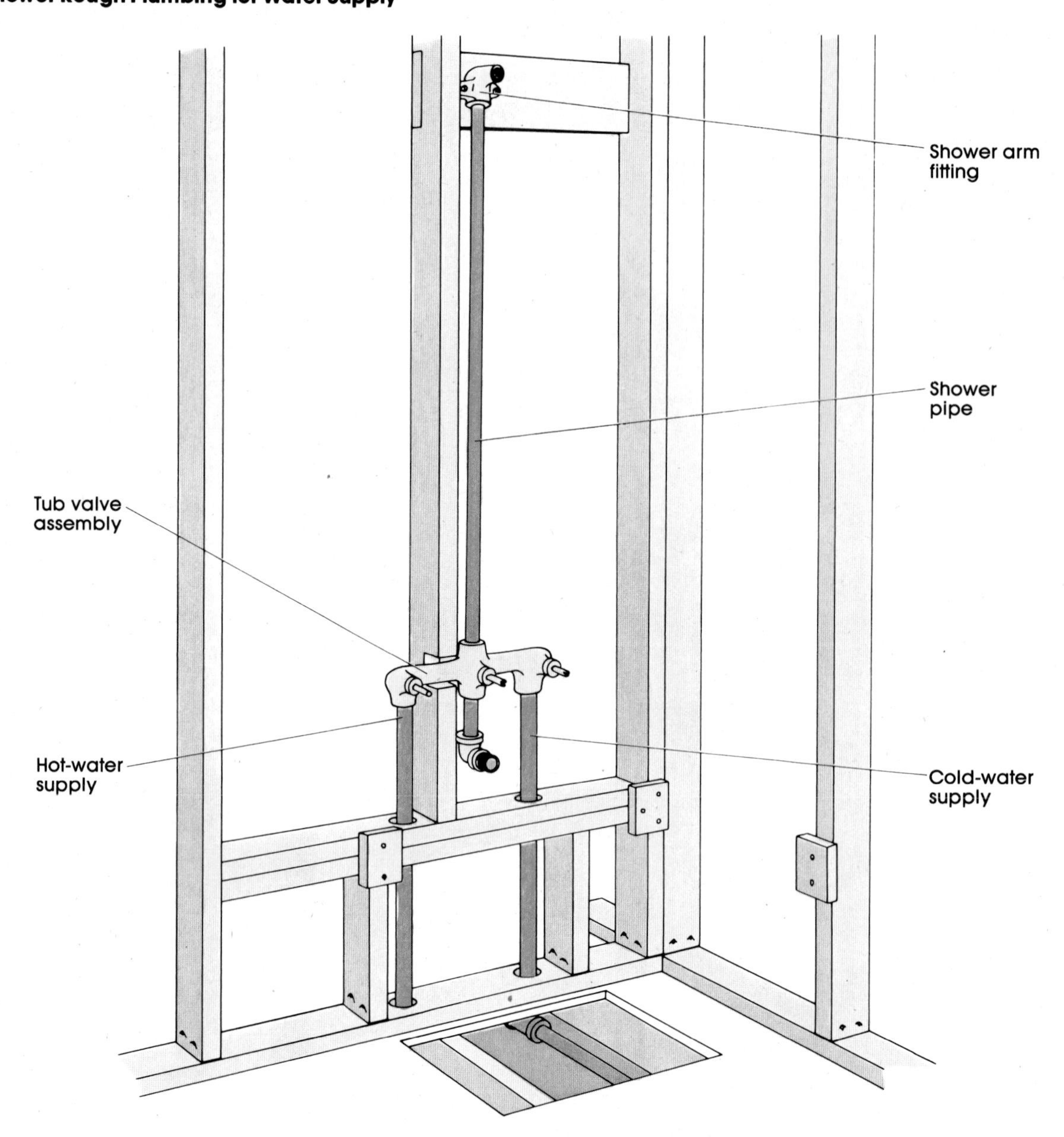

Bathtub Drain Arrangement

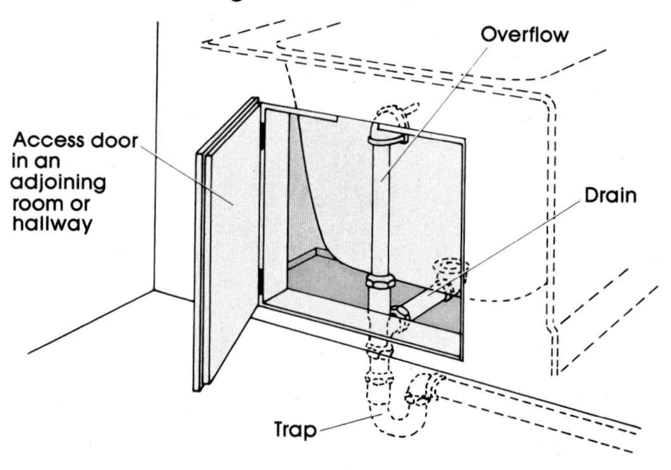

flanges, if there are any, rest on the cleats and secure them with screws through the holes in the flanges. Next you can connect the drain.

If you are making a new or drastically changed installation, be sure to install a trap (see page 67) and be sure the tub's overflow connects with the tub drain before the trap, not after. Use lots of plumber's putty to seal the tub drain. Put it under the rim of the drain in the tub and under the tub. Do not tighten the compression nut too hard or it may crack the porcelain; just make it snug. See the drawing for a typical drain installation. Finally, slip the drain plug mechanism into place, and that's it. Once you refinish the walls and floor, you will have completed the project.

SHOWER STALLS

If a shower stall needs repairing, it is usually a matter of cracked tile, loose grout, or a leaky shower pan. The tile and grout are fairly simple to take care of, as discussed on page 65. The leaky shower pan is unfortunately not a problem you can usually repair. When you finally discover that a shower pan leaks, it has most likely been leaking at least a little bit for a long time. The leaking water has probably damaged the wood structure supporting the shower. The only successful way to fix it is to tear out, inspect, and repair all the underpinnings and start over.

When the usual way to make a shower stall floor was with tile, building codes and inspectors were very demanding in their requirements because water standing on a tile surface will always seep through eventually. Lead, copper, or elaborate hot-mopped shower pans were required, and they were given strict tests before being approved. If you want to install a tile floor in your shower, contact your local building department for all the regulations.

With the advent of molded plastics, fiberglass, and steel shower stalls and sturdy plastic bases even for tile showers, a secondary waterproof shower pan became unnecessary,and it is no longer required by most codes. Because each manufacturer makes these shower stalls in its own way, the installation instructions will vary. Be sure to read and follow to the letter the step-by-step instructions that come with the shower you choose. Be especially careful to make all measurements accurately and all the framing square and plumb.

The supply piping and drain lines for the showers must also be accurately placed. Use flexible copper tubing for the supply lines so minor adjustments can be made at the time of final placement of the stall and the pipes.

Tile showers, too, have become relatively easy to install. Select a plastic shower pan that is the size of the shower you want and frame the stall to fit it exactly. Set the pan in the bottom and connect it to the drain line as shown. Cover the framing with waterproof (green) sheetrock, being sure the sheetrock covers the lip at the top of the pan to within ¼ inch of the curb. Cover the sheetrock to a height of at least 6 feet with the tile of your choice, and you have a shower stall. Setting the tile in a waterproof mastic and applying the grout is a rather easy do-it-yourself project. Your local tile dealer will be delighted to give you instructions.

Tile and Fiberglass Showers

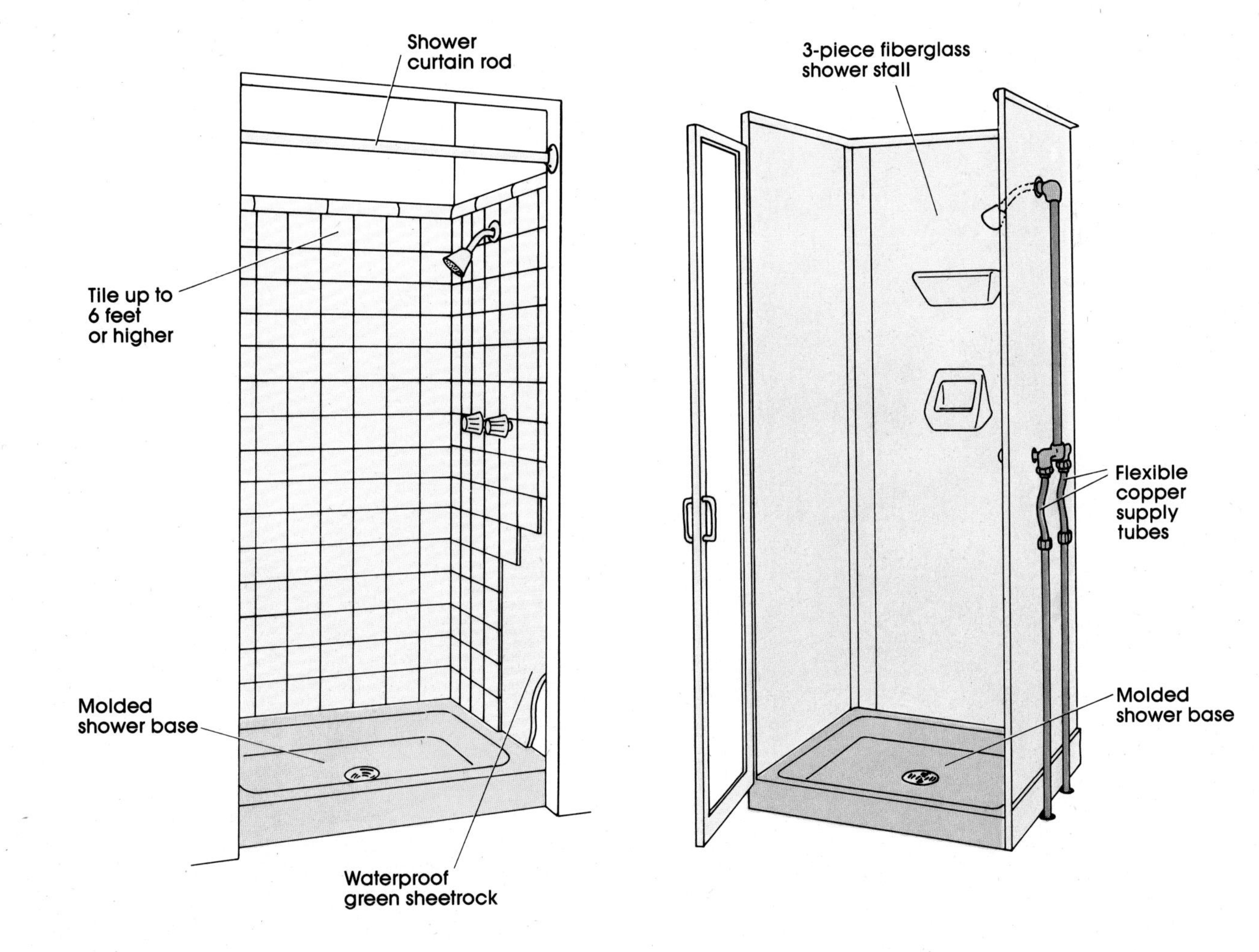

KITCHEN SINKS

What most often needs repair on a kitchen sink are chips in the enamel finish, leaky faucets, leaky drain lines, and leaky or defective strainers. See page 52 for leaky faucets and page 67 for P-traps. The repair of chips in the enamel finish can be accomplished with matching an appliance touch-up enamel available at your hardware store.

Replacing a Sink Strainer

If your sink strainer either will no longer hold water in the sink or it leaks water under the sink, it's time for replacement. If the metal is corroded or any rubber parts have deteriorated, you should replace the stopper basket or the whole unit. If it's leaking water under the sink, you may only have to replace the gasket or bead of putty between the body of the strainer and the sink bottom. In either case, here's how.

Check under the sink to see if your strainer is held by three screws in a retainer or by a single large locknut on the strainer. Loosen the slip nut holding the tailpiece between the sink and P-trap and slide it down onto the P-trap. Remove the retainer screws or loosen the locknut by turning counterclockwise. If you don't have a pipe wrench big enough, drive the locknut open with a hammer and a piece of wood wedged into one of the notches on the nut. With a screwdriver, pry the strainer out from the top. When reinstalling or putting in the replacement, make sure the sink is completely dry. Coat the underlip of the strainer with a thick bead of plumber's putty or install a new gasket. Press the strainer firmly into place and then replace the retainer or nut underneath. In tightening the nut, the strainer can be held in place by slipping the handles of a pair of pliers through the strainer and putting a screwdriver between the handles to apply counter force.

Installing a New Sink

Installing a new kitchen sink can be a very easy matter

Kitchen Sink Installation

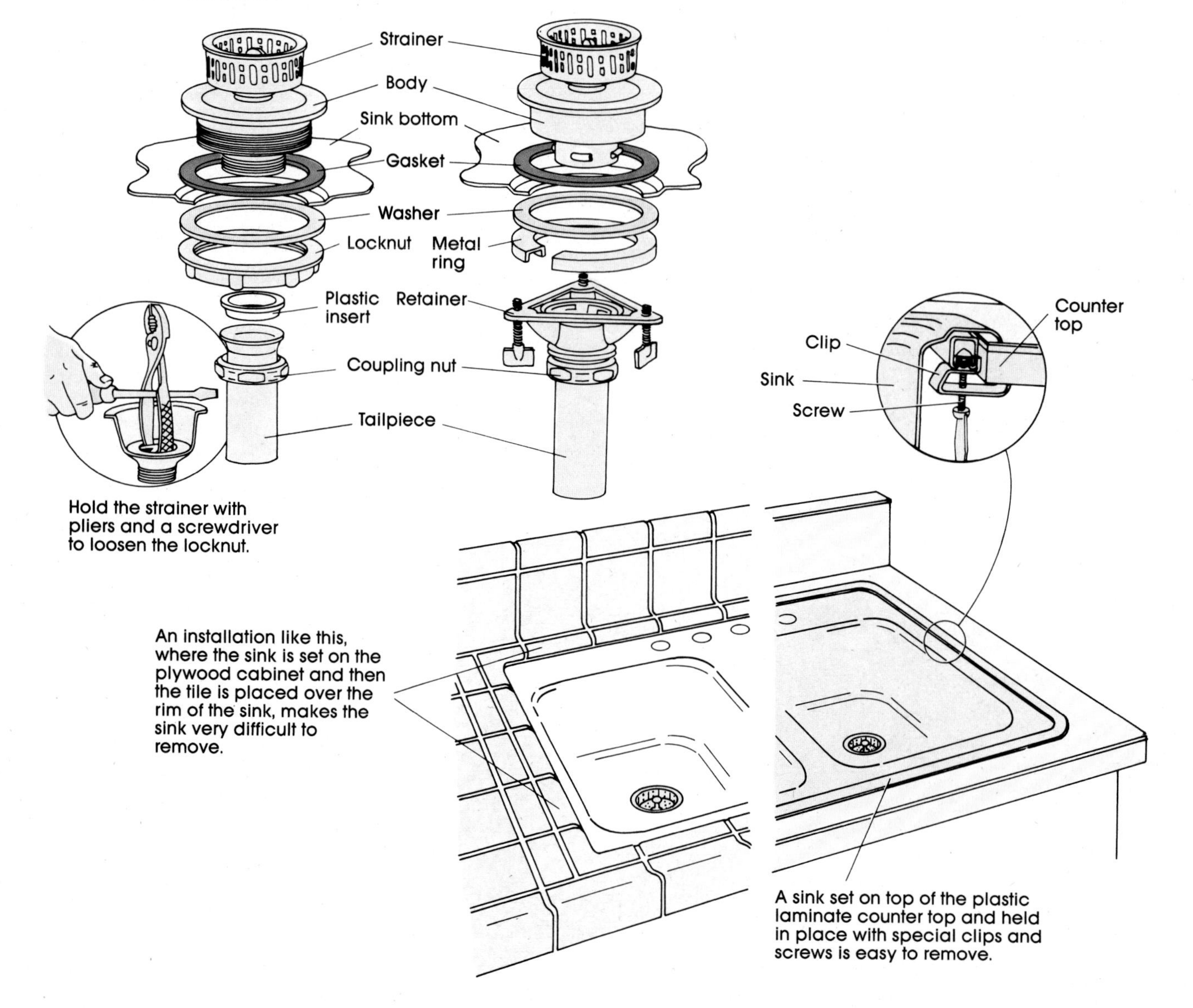

Hold the strainer with pliers and a screwdriver to loosen the locknut.

An installation like this, where the sink is set on the plywood cabinet and then the tile is placed over the rim of the sink, makes the sink very difficult to remove.

A sink set on top of the plastic laminate counter top and held in place with special clips and screws is easy to remove.

or a messy one, depending on how your old one is set on the counter. If your old sink is self-rimming and is set on top of the tile or plastic laminate or if it has a stainless steel rim around it, you have a fairly easy job.

Any tile or plastic laminate counter top on top of the sink's rim will have to be removed before you can remove the old sink. This means you will have to replace the counter top as well as the sink. If you must do this, be sure to replace the counter top before you install the new sink so the same problem doesn't occur in the future.

Installing a new kitchen sink is like installing a new lavatory (see page 65) with a few exceptions. Most kitchen sinks have two compartments that drain into the same line (see the drawing for the various ways to make this connection). The faucet assemblies for kitchen sinks are a little different, too. They are usually larger than bathroom faucet assemblies and often have a spray hose in addition to the faucets. See the drawing for how these are assembled.

Ways to Connect a Double Sink to a Single Trap

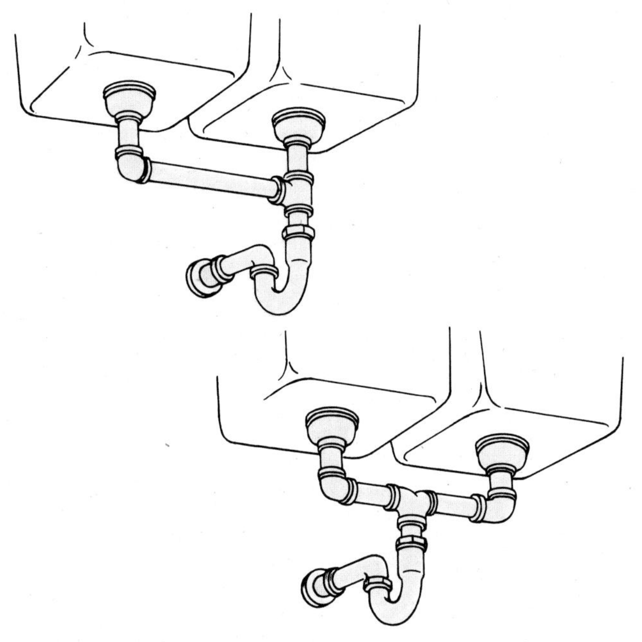

Faucet Assembly

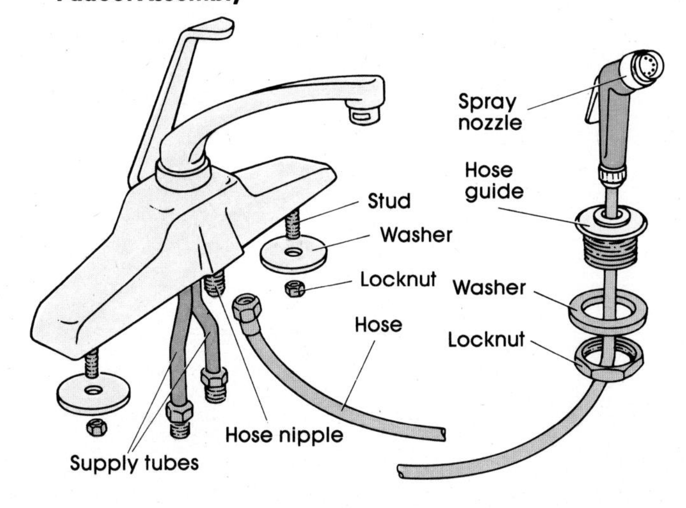

Garbage Disposer Drain Installation

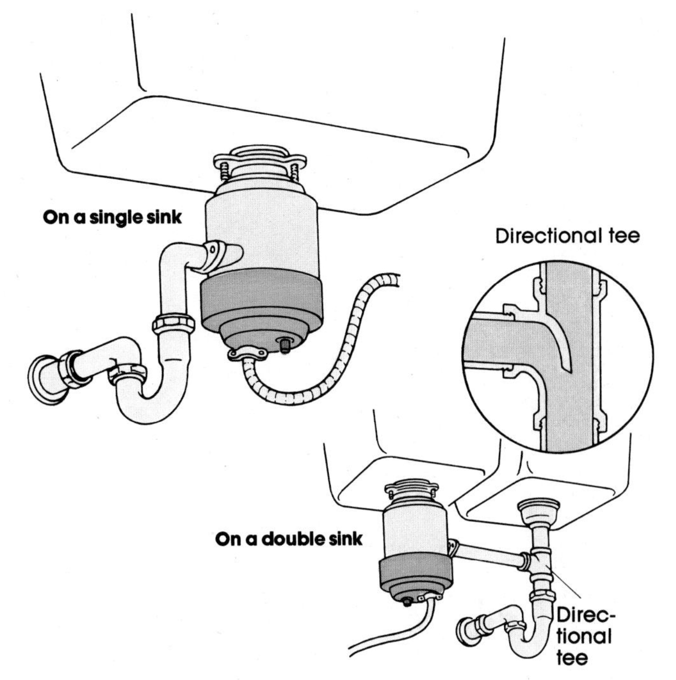

Garbage disposers are fairly controversial as far as building and plumbing codes are concerned. When they first became popular in the early sixties, almost all new houses were equipped with them. In recent years, expanding populations in many areas overloaded marginal sewage disposal facilities and the additional discharge of garbage into the system became intolerable. Local codes in these areas forbid the installation of disposers. In other areas, where sewage disposal systems are adequate, some local authorities feel that disposers are more efficient in handling this waste than the garbage collection system, and they require their installation in new construction and major kitchen remodeling. Be sure to check the situation in your area at the planning stage.

Garbage disposers are designed to fit the standard 3½- or 4-inch drain outlet of kitchen sinks, and they require a nearby 120-volt grounded electric outlet. Units from various manufacturers differ greatly. However, each comes with detailed installation instructions, which you should study carefully and follow exactly.

■ CAUTION: The combination of water and electricity is very dangerous. Always be sure the electricity is shut off when you do any work on an appliance that connects to both the electrical and plumbing systems. Be sure everything is exactly as it should be, especially the ground connections, before you turn the electricity back on.

Some codes that allow waste disposers require separate trap and drain pipes when you have a two-compartment sink. Others allow the disposer and the other sink to discharge into the same trap as long as a directional tee is used where they connect (see the drawing). If your sink only has a single compartment, you can install the disposer onto the existing trap. There are usually no additional code requirements.

DISHWASHERS & WASHING MACHINES

A dishwasher is another of those appliances which connect to both the electrical and plumbing systems and pose a great danger of shock. It requires a nearby grounded electrical connection or receptacle.

Dishwashers come in two basic styles, portable and built-in. Some of the portable ones are designed so they can be built in at a later date if the owner wants. These are especially good for renters who are planning to buy a house in the near future. The portable units require no plumbing. The supply hose just snaps onto the kitchen faucet, and the drain line is simply hung over the edge of the sink. Because no permanent connections are made, no permit is required and code requirements do not apply. A built-in unit does not necessarily require a permit, but its installation is often covered by some rather stringent code requirements (see page 7).

Repairing a Dishwasher

The basic maintenance procedure on a dishwasher involves cleaning the water jets on the sprayer. Unscrew the nut holding it, lift the sprayer out, and use a piece of wire to pick out any food particles wedged in the holes.

If the washer will not run at all, check the door. The door must close properly and tightly to close the door switch. Wait a few moments. The machine may just be between cycles. Also make sure nothing is jammed against the spray arm. If it still doesn't work, it is not a plumbing problem. Call a professional to fix it.

If the dishwasher overfills, there are a couple of things to check. Most machines are protected against overfilling by a float switch that closes the solenoid valves when the proper water level is reached. A float connected to a switch rises and falls with the water level. Jiggle the float, which is usually located in a front corner of the tub, to be sure its arm is free. If the arm is free and the switch still doesn't work, it is probably defective. Buy a new one at your appliance dealer and replace it.

If your washer won't drain, what you do depends on how the washer empties itself. Some are drained by the motor reversing itself and pumping the water out. Others just have a drain in the bottom, and the water flows out by gravity. If yours has just a gravity drain, find it and clean it out. For a reversible motor pump, you'd better call a repair person.

Dishwasher leaks are usually either in the gasket around the door or in the water supply hose. If the gasket is cracked or broken, replace it. If the hose leaks, check the clamps first and tighten them. If it still leaks, replace the hose.

If the cycle and timer switches seem to be faulty, there is usually nothing you can do without a mechanical inclination and a service manual for that particular washer. It might be better to call a repair person.

Installing a Dishwasher

Installing a built-in dishwasher is not very difficult. All dishwashers come with detailed instruction sheets. Some manufacturers make theirs especially simple and even illustrate them for the do-it-yourself homeowner. Some municipalities require a permit and an inspection to install a built-in dishwasher, so check out your local regulations.

The space required is that of a standard kitchen cabinet: the opening must be 24 inches deep, at least 24 inches wide, and 34½ inches high. The opening must be square and plumb. A dishwasher will not work properly if the frame is distorted or not level. In addition to the cabinet opening, you must prepare the electrical, hot-water supply, and drain connections before installation. The electrical connection must be a grounded junction box into which the dishwasher can be wired or a receptacle (outlet) for a three-prong plug.

Most dishwashers are connected to the hot-water

Parts of a Dishwasher

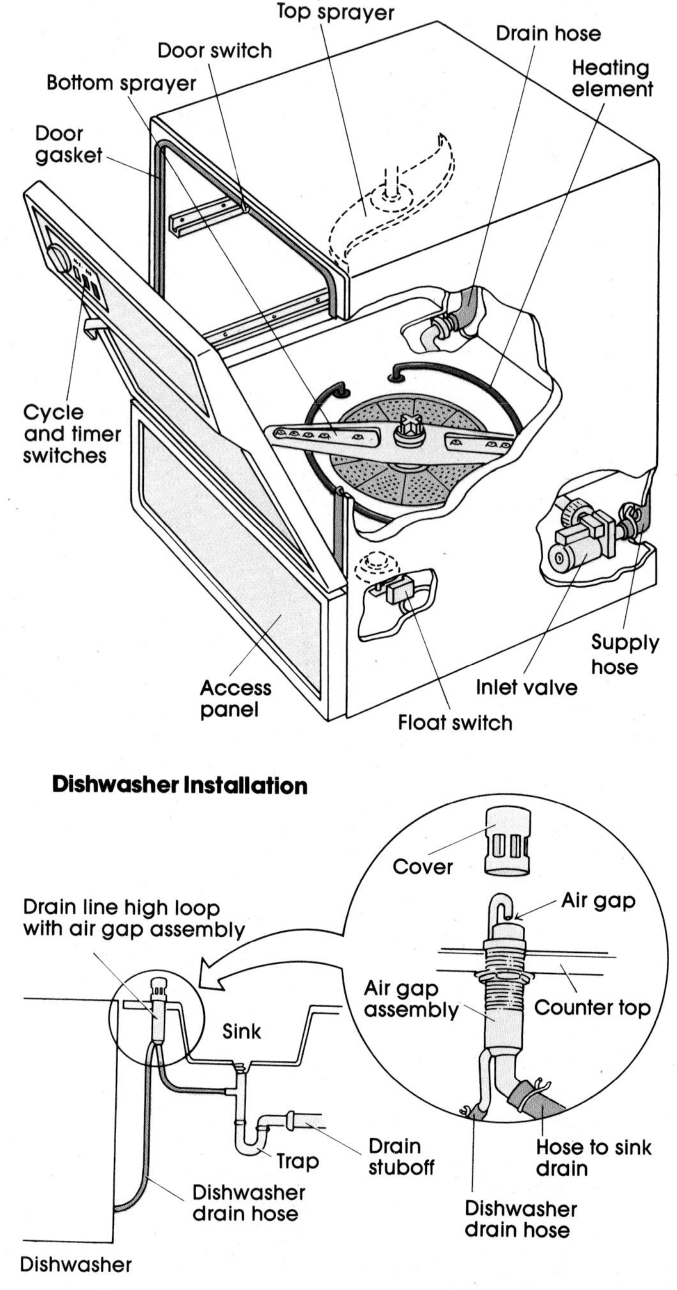

supply line and should have their own supply stop even though there is a built-in valve near the water inlet. Dishwashers also need to be connected to the kitchen drain line. Any units that have a pump to discharge wastewater must also have a drain line that rises to the height of the top of the washer. This *high loop* prevents a back-up of wastewater into the dishwasher if the sink gets clogged. Some codes require an *air gap* fitting on this high loop to prevent siphoning. Even if it's not required by your local code, you should consider installing one. The air gap assembly mounts on the counter above the rim of the sink. Some sinks have an extra hole near the faucet assembly for this purpose (see the drawing). The drain line should also be connected to a special "dishwasher tee" that you have installed on the tailpiece of the kitchen sink or, if you have a garbage disposer, to the special opening on the side of that unit.

Installing a Washing Machine

A washing machine is always a free-standing appliance that can be added, moved, or replaced at any time. Most building codes now require that new living units be provided with all the connections necessary for the installation of a washing machine. All you need to do if you live in one of these units is screw on two hoses, connect a drain hose, and plug the machine in.

If you live in an older house that does not have these connections, they are relatively easy to install yourself. Look at the drawing to see what is needed and refer to page 54 for how to do it. The standpipe and air chambers or shock absorbers are very important in this installation. Don't leave them out.

Washers have solenoid valves to turn the water on and off. These valves open and close instantly, not gradually like a faucet. Air chambers or shock absorbers on both the hot- and cold-water pipes minimize noisy and destructive water hammer when these valves operate. If possible, install the hose bibs where they can be reached when the washer is in place. The hoses and solenoid valves will last longer if the water is turned off between washing days.

The standpipe provides protection from back-up and siphoning like an air-gap fitting does on a dishwasher. Prefabricated 2-inch standpipes are available from your dealer in several lengths. They have a built-in trap and are designed to fit into a standard 2-inch drain pipe.

When you have made all the connections and your washer is in its permanent position, it must be leveled. Check it from front to back and from side to side with a carpenter's level. Threaded pads can be turned to raise or lower each corner of the washer.

The only kinds of trouble we can deal with here are those involving the plumbing, that is, the hoses and fittings that connect the washer to the water supply and drain lines. If the trouble is within the machine, you will have to consult another book or call a professional repair person.

If the washer won't fill or fills very slowly, check the supply stops in back of the washer. Someone may have turned them off. Those valves must be open all the way

Washing Machine Connections

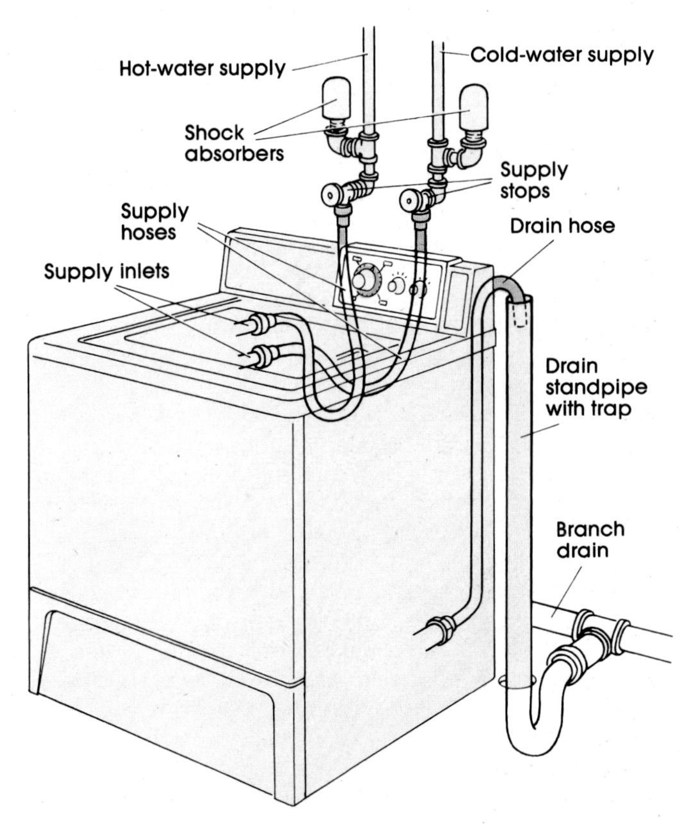

for the washer to fill quickly. If the valves are on, but the trouble persists, check the hoses. Be sure they have not been bent too sharply, kinked, or squashed between the washer and the wall or a pipe. There is one more thing to check if the hoses are all right. Turn the supply stops off and unscrew the hoses from the washer. Just inside the hot- and cold-water inlets you will probably find screens, which keep sediment and water pipe debris out of the washer. If the washer has been there for a long time or if you've been working on the plumbing system, they could be clogged. Clean the screens or replace them. If everything is clear this far and you still have a problem, it has to be inside the washer—probably a solenoid valve. Check a repair manual or call a repair person.

Sometimes a washer will drain when it is not supposed to. If it was installed without a stand-pipe so the drain hose never rises higher than the water level in the washer, it is possible that the water is being siphoned out. Obviously you need to put in a standpipe or at least raise the hose so it is higher at some point than the washer's water level.

If you find water leaking from the washer, check all the hoses and connections in back of the washer. If a hose connection has vibrated loose, tighten it. Replace any old and leaking hoses. If a supply stop has a stem leak, see page 53 for how to replace the packing. If the leak is from the washer itself, it is probably from a pump, the drain valve, or the tub seals. Check a repair book or call a repair person.

Your water heater is one of the biggest energy users in the house. Only the furnace or central air conditioner uses more. Proper care and maintenance will save you money, energy, and grief by keeping it working efficiently and smoothly for a long time.

Make it a practice to drain a bucket of water from the drain spigot at the bottom of the heater every couple of months. This keeps sediment from accumulating in the bottom and reducing the heating efficiency.

At least once a year, test the temperature/pressure relief valve by pulling up or pushing down on the handle. If hot water spurts out the overflow pipe, it is working. If your heater doesn't have a temperature/pressure relief valve or an overflow pipe, it is imperative that you install one (see the drawing).

All the late-model water heaters have an energy cutoff device that automatically stops the power or fuel to the unit if the water becomes dangerously hot. It prevents the water from boiling, building up steam, and possibly bursting the tank. If you find that the water has become hotter than you want it and turning down the thermostat doesn't make it any cooler, you probably have a defective thermostat. Call a professional to replace it.

A rumbling sound in the heater is another indication that a defective thermostat may be allowing the water to get too hot. Check the thermostat to see that it didn't accidentally get pushed to the maximum. If not, turn the heater off immediately and call a professional.

If you begin to get rusty water from the hot-water faucet and not the cold, it indicates corrosion in your hot-water tank. Drain the heater until the water runs clear and you've removed the sediment buildup. If there's a lot of rust, your heater probably has terminal corrosion and you should start budgeting for a new water heater.

Temperature/Pressure Relief Valve on a Water Heater

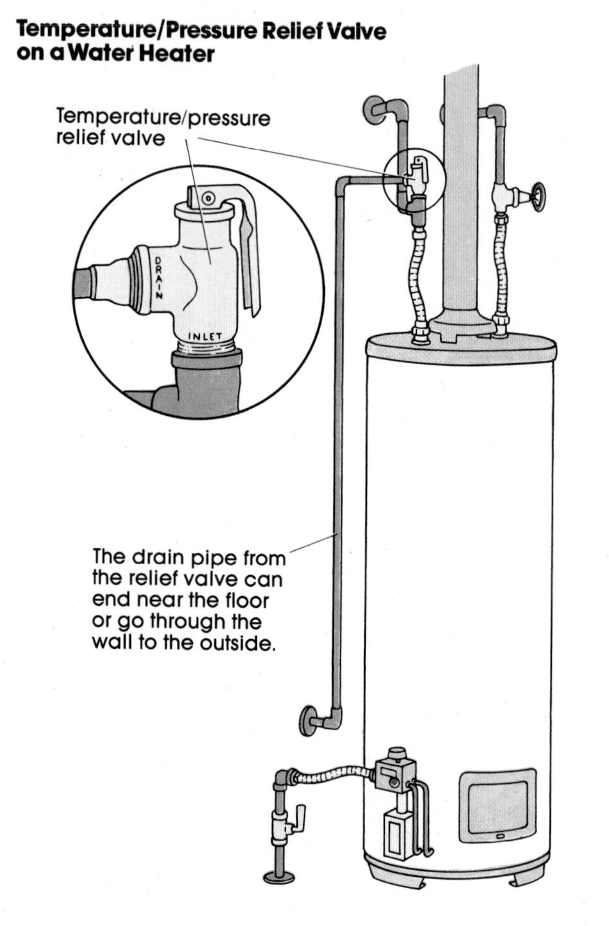

Installing a New Water Heater

If you are installing a water heater in a new location rather than where your old heater was, here are a few things to consider. You can put an electric water heater almost anywhere as long as there is an approved power supply and a place for water to drain should it leak or overheat. Building codes prohibit installing a gas water heater in an occupied room that is normally kept closed to outside air, such as a bathroom or inside laundry room. Put the heater as close as you can to the bathroom or laundry where the hot water will be used. Don't put the heater in an area exposed to outdoors where it could freeze when it is not being used.

If a gas water heater is installed in a closet or other enclosed space, it must be ventilated as prescribed by the building code. Codes usually recommend two vent openings, one at the top and one at the bottom of the door or wall. Each opening must have a minimum area of 100 square inches with a width to height ratio of two to one. The building code also prescribes minimum clearances between the water heater and combustible materials. These are usually 1 inch at the sides and back; 6 inches in front; and 18 inches from the top of the tank. Codes usually stipulate that a gas water heater located in a garage must be raised 18 inches off the floor to prevent leaking gasoline or oil fumes from contacting the pilot light.

When planning the piping for a water heater, remember that the cold-water supply pipe to the heater must be at least ¾-inch pipe, and it must have a shutoff valve in a easily accessible location near the heater. The pipe carrying hot water from the heater may be as small as ½-inch but ¾-inch is probably desirable. Both the hot- and cold-water pipes should have a flexible hose or union near where they connect to the heater.

If you are installing a gas heater, be sure the gas piping meets all code and gas company requirements. The piping must have an easily accessible shutoff valve nearby and be connected to the heater with an approved flexible connector.

When you have your new water heater and are ready to put it in, shut off the water and fuel supply to the old heater. Drain the tank completely through the spigot at the bottom and then disconnect the unions or flex hoses that go to the hot- and cold-water pipes. On oil or gas-fueled heaters, remove the draft deflector or collar that links it to the flue pipe.

Unless you've had some indication that it is defective, the temperature/pressure relief valve is still good. Check the manufacturer's recommendations, and if the old valve is compatible with the new heater, remove the overflow pipe and the valve to use on the new heater.

Remove the old heater and set the new one in place. If the pipe fittings on the new heater don't line up with

How to Install a Water Heater

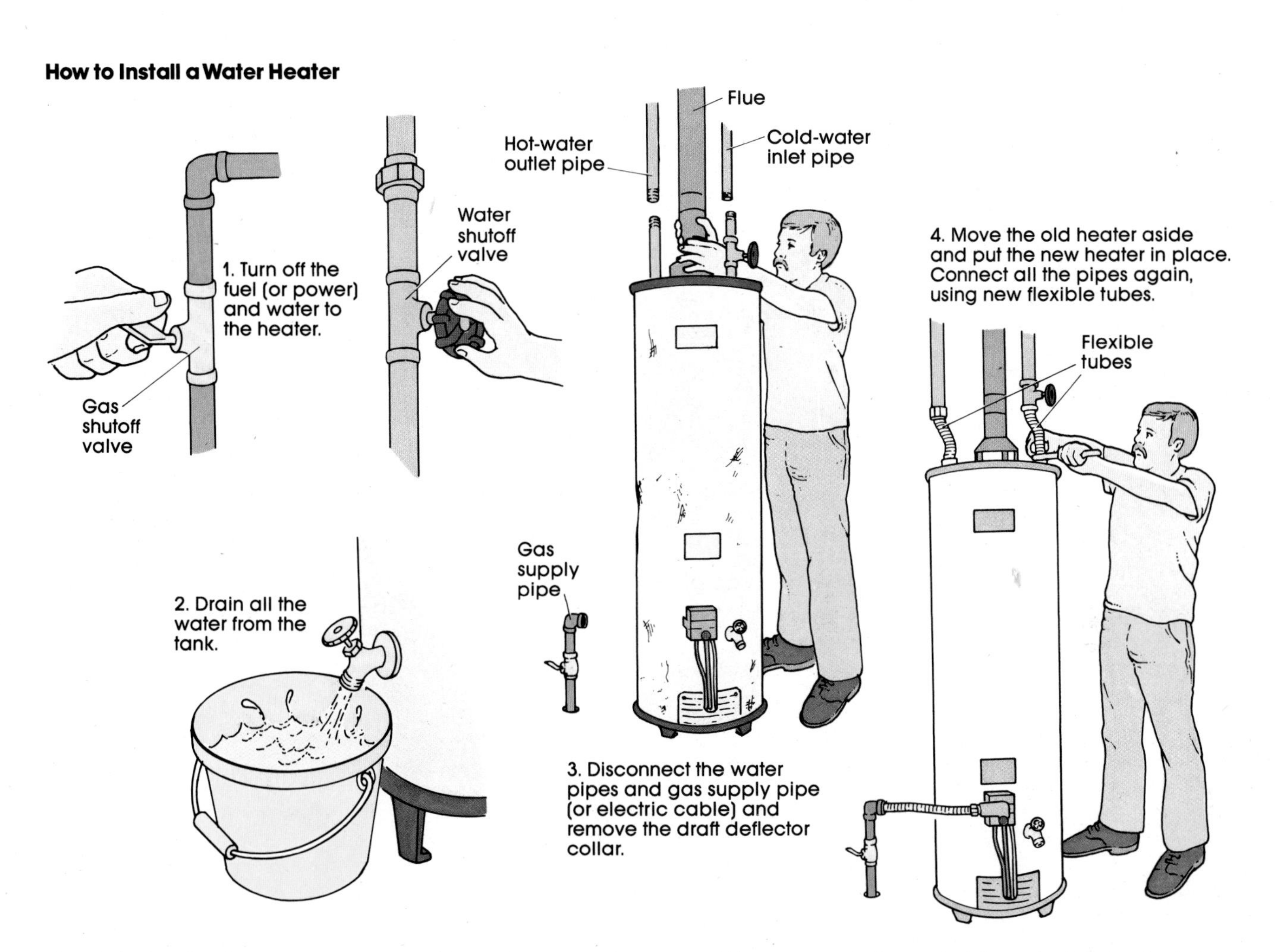

the old pipe hookups, you may want to add a pipe nipple or two. Or if the new heater is too low, you can put some stout wood blocks or bricks under the feet to raise it. The best way to hook up both the water and gas is with flex-pipe connectors. Unless there is a drastic change in the configuration of a new heater, you just bend the connectors a little to fit; you don't need to add or subtract pipe or fittings.

When all the connections are made, turn on all the valves. When the tank is full of water, turn the power supply back on, and it's ready to go. With a gas heater, use soapy water to check all the gas connections for leaks and light the pilot.

Insulation

To save energy and money, your water heater and at least some of your hot-water pipe should be insulated. If the storage tank of your water heater feels warm to the touch, it is losing heat. If it's in a closet or laundry room within the area of your house you heat during the winter, this is not too bad because the lost heat is helping the furnace by contributing to the space heating within your house. On a hot summer day, however, you may not appreciate the extra heat, especially since you are paying for it. If the water heater is in an unheated basement or on an open porch, it is really losing a lot of heat and wasting a lot of energy.

All commercially available water heater tanks have some built-in insulation. Electric heaters usually have about 2 inches of insulation, and oil and gas-fueled heaters commnly have about 1 inch. These heaters tend to lose from 20 to 35 percent of their heat through the walls of the tank. If they are in a location exposed to cold outside air, it can be quite a bit more. This amount of heat loss can account for $2 to $3 of your gas or electric bill each month.

A single roll of 3½-inch thick R-11 fiberglass insulation and a roll of duct tape to attach it will cost about $20 and will be more than sufficient to wrap a 60-gallon water heater. This $20 investment will usually pay for itself in about a year and save you both money and energy for many years to come. You can buy precut kits or preformed insulating jackets to fit most water heaters, but the job is so simple that most people don't consider them worth the extra cost.

If you are going to insulate your heater, make sure it is equipped with a temperature/pressure relief valve. The additional insulation may allow the water to overheat if you don't use any hot water for a while. Most state and local codes require these valves. Even if yours doesn't, it's still a very good idea to have one on your heater.

When insulating a gas or oil-fired water heater, don't block the pilot light or access to any controls. Do not put the insulation where it's exposed to the burner flame or

Insulating a Water Heater

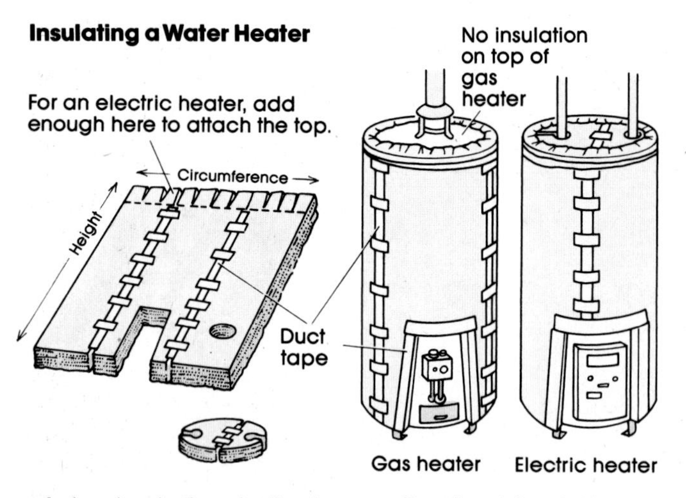

obstructs air flow to the burner. Don't put insulation on the top or near the flue pipe. On an electric heater, you may put insulation on the top, but do not cover the control panel. Before you cover the name plate on your heater, make a note of the name, model number, and serial number just in case you need to call for service or want to order a new part for it.

When insulating your water heater, you may as well spend a little more time and money and insulate some of the pipe, too. Insulate at least the first 10 feet of the hot- and cold-water lines connected to the heater. If your heater is in an unheated area, insulate the pipe until it goes into the heated part of the house. This will not only save fuel and money, but will minimize the length of time you will have to wait for hot water to come out of the faucet after you turn it on. To insulate the pipes, you can use preslit foam sleeves or adhesive-backed foam tape that is made for the purpose. Or you can cover your pipes with a combination of ½-inch thick vinyl foam and duct tape.

For more information about insulation and other ways to save energy see the Ortho book *Energy-Saving Projects for the Home.*

Solar Water Heating

If you are thinking of doing some energy saving in your home with solar power, water heating might be the best place to start. Even at its most complex level, solar water heating is less expensive than solar space heating, and since hot water is needed all year around, the cost is amortized more rapidly. A simple nonstorage system is impractical for most households. The average daily consumption of hot water in the United States is about 10 to 15 gallons per person. This use includes washing machines, dishwashers, and bathtubs, which consume a disproportionately high percentage of our hot water and operate principally in the morning and evening hours when nonstorage solar water heaters are not effective.

As with any other solar energy adaptations for the house, conservation should be your first step because it requires the least effort or expenditure in proportion to the savings it returns in both energy and money. Less use translates into a need for smaller storage capacity and smaller collector area relative to the proportion of hot water supplied by the system. By the same token, greater use means the need for greater collection ability and facilities for storing more water or storing it for a greater period of time. With solar-heated water, the energy savings are reflected directly and immediately in the size, complexity, and cost of the system you install.

The best kind of water heating system for you, of course, depends on your particular circumstances. A useful rule of thumb to remember is that in the United States, a well-engineered solar water heating system, installed and used properly, can supply a typical household with close to 100 percent of its summertime hot-water requirements, about 25 to 30 percent of its winter requirements, and between 50 and 75 percent of its annual hot-water needs. The cost of a solar hot-water system will run in the vicinity of $2,500 to $3,000, installed. Your eventual savings will depend on the way you use water and the cost of your present water-heating system.

Passive systems. The simplest form of a practical solar water heater is called a *bread box collector*. The solar heat is collected and stored in the same unit. There is no transfer from one part of the system to another. However, all storage heaters lose at night the heat they collect during the day. Storage heaters are essentially passive systems except at the use or distribution stage. An elementary mechanism, like a faucet, allows user control.

The bread box has been around in one form or another for about half a century, but its basic design remains the same. The storage heater is usually located above the water outlets. This way gravity does all the work and the system is kept as simple and inexpensive as possible. It has one or more tanks, which are painted black, contained in an insulated box covered at the top and usually on the south face by double or triple glazing. The box is opened, generally by hand, in the morning to expose the glass and tank to the sun. It is closed at night to minimize the loss of heat. The tank can be filled by some independent method, such as a hose, or can be connected to the main water supply of the house. The insulated, solar-heated water in the tank is unlikely to freeze in most areas. However, if you live in an area subject to deep freezes, you probably shouldn't use this system. In any case, insulate the pipes that carry water from and to the heater. This will minimize ordinary heat loss and guard against freezing.

Solar heat can be collected at a point below the storage tank and transported to the tank by thermosiphoning. This action relies on gravity, taking advantage of the fact that warm liquid rises and cooler liquid falls. In *thermosiphon systems,* cool water is warmed by the solar collector at the bottom of the system. It rises and is piped to the top of the storage tank. As it cools, the water falls to the bottom of the tank and reenters the collector. Water for use in the house is tapped from the top of the storage tank, where it is warmest. The continuous cycle raises the temperature of the water in the storage tank as long as the sun shines on the collector. How long the solar-heated water remains warm after the sun ceases to shine depends on the capacity of the storage tank and how effectively the tank is insulated.

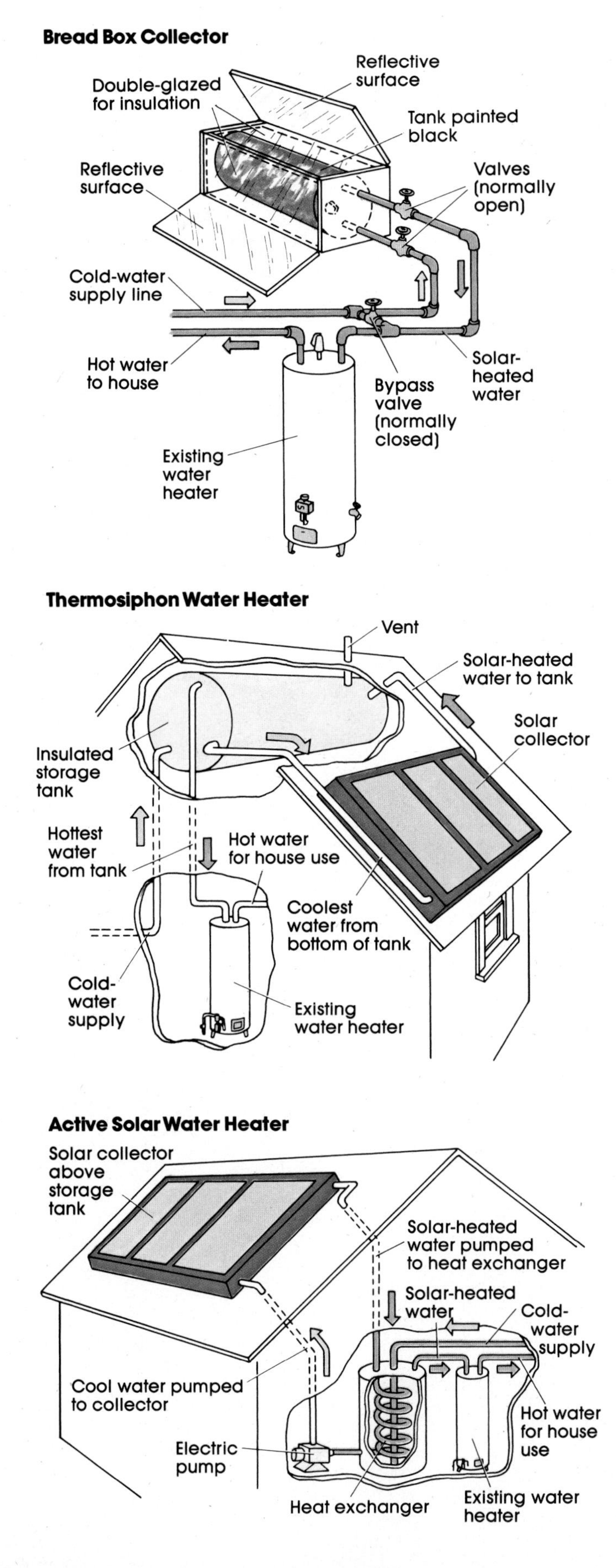

Active systems. Most solar hot-water systems available commercially are active systems, which use pumps to circulate the water. They are also equipped with controllers that turn on the pumps only when the sun is heating the water in the system.

The pumps most commonly used are 1/12 to 1/36 horsepower depending on the number of collectors, size of pipes, and height difference between the storage tank and the collector. The controllers used to activate the system measure the temperature difference between the coldest part of the water storage tank and the warmest part of the collector. The pump is turned off or on by signals from the controller. To activate the pump, the collector needs to be about seven degrees (F) warmer than the storage tank. This temperature difference means there is heat to be gained in the collector, and the water is moved to retrieve it. As long as the temperature difference remains, the pump will stay on and the water will be sent to the collectors to continue the cycle. When the sun sets or is covered by clouds, the water in the collector becomes cooler, and the controller shuts down the pump until the sun again comes out, and the collector becomes warmer than the storage tank.

Auxiliary systems. Auxiliary solar water-heating systems are connected to an existing hot-water system. The water supply that would normally enter your water heater as cold water is first piped into the solar heating system. When you turn on the hot-water faucet, water is drawn from the solar storage tank into your gas or electric water heater. The heater's thermostat has been set for the temperature that you want your hot water to be delivered so it heats the entering water to that temperature. For instance, if the water comes into your house from the water company at 60 degrees and you have your heater's thermostat set at 130 degrees, as soon as the solar system has heated the water to 100 degrees about half of your water heating has already been accomplished. Any heating done by the sun is that much less work required by gas or electricity.

Heat exchangers. If there is any possibility that freezing or near-freezing temperatures might reach your solar system, insulation is critical. It takes only one cold night to burst pipes that may require complete replacement to repair your system. One sure-fire way to prevent your solar heater pipes from freezing is to fill them with antifreeze solution. Unfortunately, you cannot drink antifreeze and live to enjoy a bath in solar-heated water.

To get around this difficulty, heat exchangers are used. The most common heat exchanger is made of copper tubes immersed in a storage tank within the house. The antifreeze solution, or transfer fluid as it's called, is pumped through the solar collector to pick up heat and then through a heat exchanger coil in contact with the potable water supply in the storage tank. The heat in the transfer fluid is picked up by the potable water in the tank, and then this heated water is piped into the gas or electric water heater as in the auxiliary system just described. There is no contact or mixing between the transfer fluid and the potable water. Whatever you do, do not allow antifreeze into your household water supply.

INSTALLING A NEW BATHROOM

With an idea, a plan, and the methods in this book, you can install the plumbing in a whole new room. You will need to size and frame pipes and fixtures and may need to cut through old walls. Here's how.

This chapter will help you to plan and install a new bathroom, either in an addition or in an existing room in your house. Because it has the most fixtures—including a toilet, which presents special problems in drainage and venting—the bathroom is the most complex plumbing installation in your house. If you can install a new bathroom or remodel an old one, which is sometimes more difficult, you will find that the steps are very similar for installing a new kitchen, laundry room, or wet bar.

Two checks of new plumbing are required by law—approval of your plans when you apply for a building permit and inspection of your work before it is covered. Not only will this inspection contribute to your health and safety, but it should also reassure you if this is your first attempt at installing plumbing.

Planning a New Installation

In your initial planning, you may want some professional help. You can either have a builder or architect do the planning for you, with the understanding that you want to do the work yourself, or you can arrange to have a professional go over your plans. Check with your local authorities to see how much detail is required in the plan you submit for your permit.

Surveying Your Plumbing System

You may have a very specific idea of how and where you want your new bathroom to be. But whether you plan to remodel or to add on a new room, the first step is to check over what you have now. Explore the plumbing systems in your house. Walk around the outside and make notes as to where pipes enter and leave. Check the roof to see where the vent pipes are. Look into the crawl spaces under the house, into the basement, and into the attic to familiarize yourself with the piping that is visible and to get a general idea of the location and size of pipes hidden within walls, floors, and ceilings.

Find where the water supply line (house main) enters the house and where along its length it would be easiest to install a tee for water supply to the new bathroom. Check your water heater to see if it will supply enough hot water for the new bathroom (see page 76) or if you will need another water heater for the new addition. If the old heater will suffice, check along the hot water main to find where it would be easiest to install a tee for the hot-water supply to the new bathroom.

Where to Tap Supply Lines

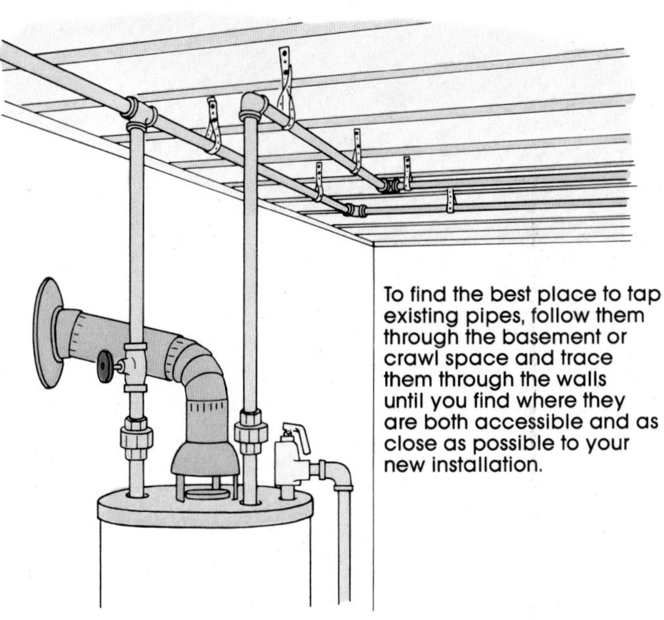

To find the best place to tap existing pipes, follow them through the basement or crawl space and trace them through the walls until you find where they are both accessible and as close as possible to your new installation.

◀

Design the kind of bathroom you really want and install it yourself.

The most important consideration in bathroom plumbing is the location and size of the existing soil stack. You will probably find it in the wall behind the toilet, in a closet, or beneath a stairway. There should be a clean-out plug at the base of the soil stack. Also note the location and size of the house drain—the large horizontal pipe that goes from the bottom of the soil stack to the sewer. A new toilet will usually drain directly into the soil stack or have its own secondary soil stack that connects to the house drain. If there is adequate crawl space, the toilet can be served by a 3-inch branch off the main drain and a 2-inch vent of its own.

The cost of the plumbing and materials and the amount of labor to install the new bathroom can be kept at a minimum by carefully planning the room's location and layout. If the fixtures in the new bathroom can be located on the opposite side of a wall of an old bathroom, back-to-back with the existing fixtures, and if the existing soil stack is of adequate size, you have an ideal situation. The amount of new plumbing will be minimal. The next-best situation is to have the new fixtures on another nearby wall and the toilet within a few feet of the existing soil stack. As long as they are vented properly, other fixtures can be quite a distance from the soil stack, but, depending on local codes, the toilet must be within 5 or 6 feet or have its own branch drain and vent.

Because most existing bathrooms are already backed up against a kitchen or another bathroom (for just the reasons mentioned here), it is unlikely that you will be able to place your new bathroom in such a position. In most cases, you will need to add a new soil stack or branch drain for the new bathroom. The easiest place to put a new soil stack is inside a new wall, as close as possible to the house drain. A new branch drain can be installed under the floor and vented through a wall.

If the new bath is in an addition to your house, the stack, branch drains, and all the new supply lines can be built into the walls and concealed beneath the floor as in any new construction. If, however, the new bathroom is going to be built within your present house, the new soil stack, as well as other drain and supply pipes, may have to be concealed by thickening existing walls, boxing them in, or hiding them in closets or cabinets. A stack can also be run up an exterior wall of a house and boxed in with siding to match the outside finish of the house. This should be done only in warm-winter areas where freezing is unlikely or if the stack will be well insulated against freezing.

Mapping Out the Bathroom

Once you have decided upon the actual location of your new bathroom, you can draw up a plan. A scale drawing is essential to accurate planning. One good method is to use graph paper, marking ¼ or ½ inch to 1 foot. Although bathrooms come in many shapes and sizes, the recommended size for a family bathroom is 6 × 8 feet and the minimum size for a full bath with tub is 5 × 7 feet.

First, draw in the room walls and mark the exact location of any doors and windows. Do not plan to run DWV or supply pipes through windows or doors. Most

Back-to-Back Bathrooms

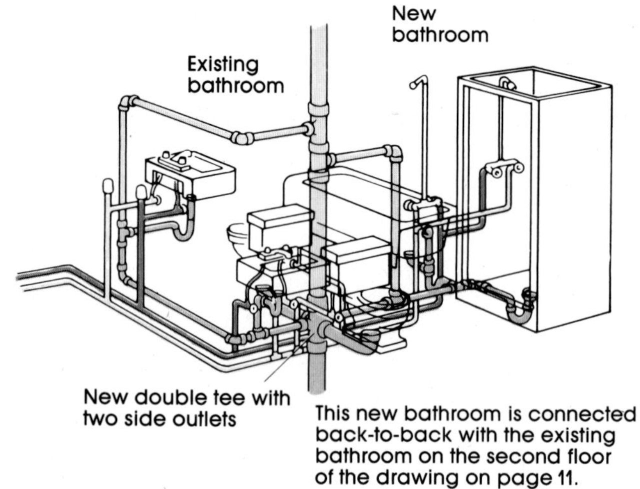

This new bathroom is connected back-to-back with the existing bathroom on the second floor of the drawing on page 11.

A New Nearby Bathroom

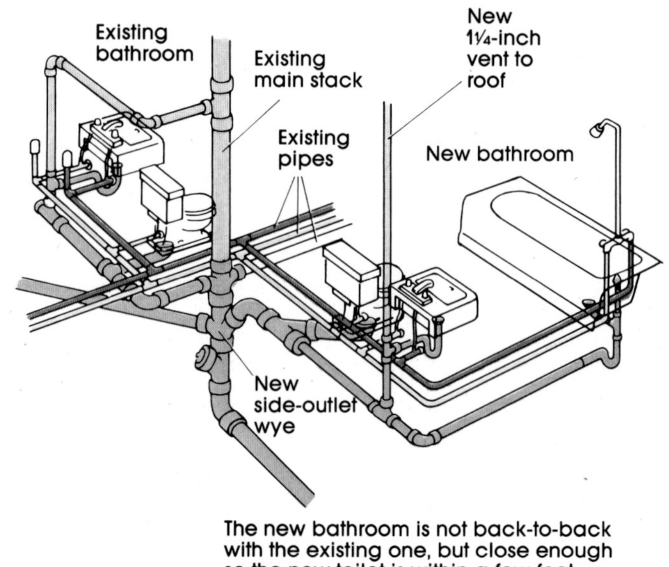

The new bathroom is not back-to-back with the existing one, but close enough so the new toilet is within a few feet of the stack.

A New Bathroom on a Branch Drain

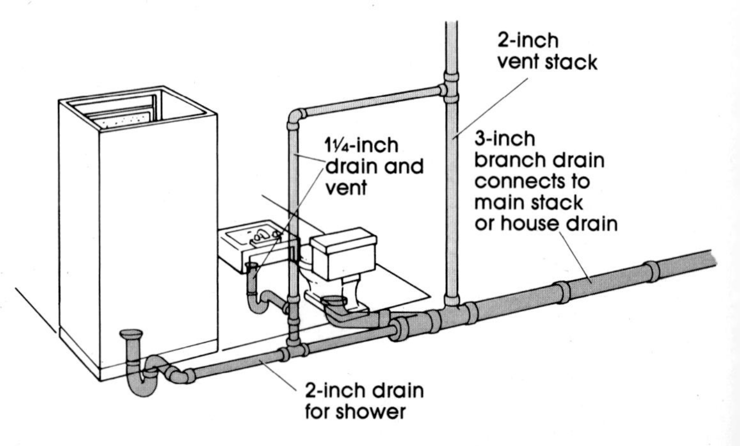

doors open into a room, but you can design the door to open out or to slide into a wall.

Next, mark the location of the soil stack. The locations of the other drain pipes and the supply pipes will be determined by the placement of the soil stack, structure of existing walls, and location of the fixtures. Pipes can be hung under first-floor joists in the basement or crawl space. Supply pipes can also be hung on top of ceiling joists in the attic. Between floors, try to plan your installation so that pipes can run the same way as the joists and lie between them.

The size and shape of bathroom fixtures can vary considerably. Once you have the measurements of your new bathroom, visit a couple of bathroom showrooms and decide on the fixtures you would like to have. To determine the exact location of your fixtures, you can cut out templates (scale models) and move them around on your drawing. One of the most important parts of planning fixture locations is allowing sufficient space for easy access, cleaning, and repairs. Although the plumbing fixture clearances in your local code may vary somewhat, you should plan for approximately the following clearances:

- *Toilet.* The minimum distance allowed from the center of the bowl to a wall or partition is 15 inches and to a bathtub, 12 inches. The minimum clearance from the front of the bowl to any wall or fixture is 21 inches.
- *Lavatory.* Allow 4 inches from the side edge of a lavatory to a toilet tank or finished wall, 2 inches to a tub, and 21 inches from its front edge to any wall or fixture.
- *Shower.* Allow 24 inches from the shower stall to any fixture or wall. The minimum floor area required for a shower is 1,024 square inches.

Fixture Clearances

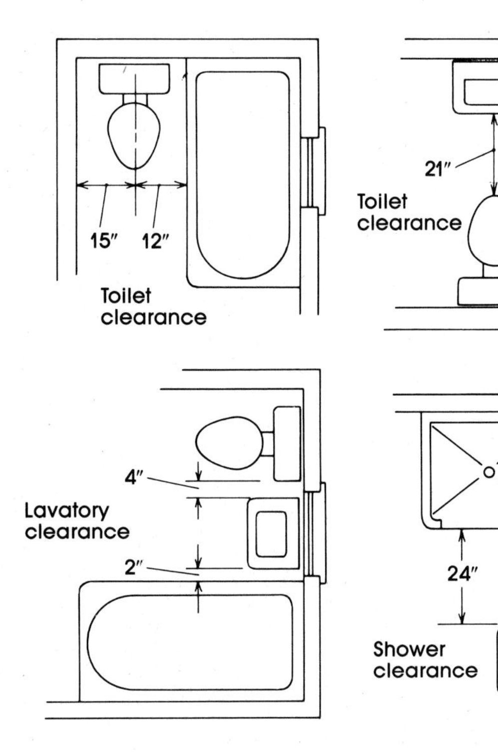

Plan the location of branch drains from each fixture. All horizontal soil pipes should slope downward about 1/4 inch per running foot. Never plan to install a major fixture, such as a toilet, upstream from a minor one. The toilet's closet bend goes directly into the stack. The lavatory, bathtub, and shower must all have traps.

Every fixture trap must be vented, either into the soil stack through the drain pipe—a process called wet venting—or through vent pipes joined to the stack above the other waste lines, which is called reventing. According to the National Plumbing Code, a tub or shower that has a trap within 3½ feet of the soil stack or a lavatory within 2½ feet of the stack can be wet vented. Check the plumbing code in your area for vent pipe specifications. The toilet will always require its own vent pipe.

Once the DWV system is planned, you can map out your hot- and cold-water supply pipes. If these will be connected to existing systems in your house, use the same route and openings as much as possible. Supply pipes should be located at least 6 inches apart and, although it is not required, can have the same slope as drain piping. On your drawing, mark the exact locations of the faucet hookups—with the cold water running to the faucet on the right and the hot to the faucet on the left.

For both the DWV and supply pipes, allow enough space for pipe insulation if you live in a cold climate. An alternative is to cut 3- to 4-inch openings in the existing foundation to allow warm air to circulate from the basement into the crawl space beneath the new bathroom.

If a new water heater is needed, decide where it will be located. For a gas heater, determine the location of its flue pipe and how the heater will be enclosed and vented (see page 76).

Fixture Venting

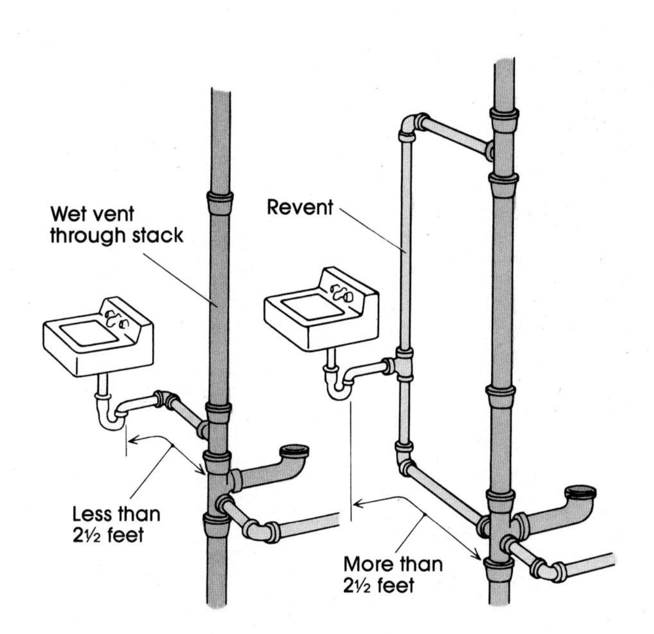

Choosing the Pipes

Carefully choose the kind and size of pipe you will use for each part of the installation. We recommend ABS pipe for all of the DWV system because it is so easy to work with. Even if the existing system is copper or cast iron, ABS pipe can be connected to it easily. The size of DWV pipe is regulated according to fixture units (one fixture unit represents a waste flow of one cubic foot per minute). The soil stack will be 3 to 4 inches in diameter and most branch drains 1½ or 2 inches. Although the table on this page lists pipe sizes according to the National Plumbing Code, you should check your local codes as well. Use the smallest size your code will allow. The code also sets a maximum length for drain pipe to run from a fixture to a stack, depending on its diameter:

DWV Pipe Length Limits

Diameter	Distance to stack
1½″	4½′
2″	5′
3″	6′

The size of the vent piping is determined by the maximum fixture unit load, the length of pipe, type of fixture, and diameter of the soil or waste stack the vent pipe serves. Again, check your local codes for specifics. As a rule of thumb, the diameter of a vent cannot exceed the soil stack diameter. The vent diameter cannot be less than 1¼ inch or less than half the diameter of the drain it serves, whichever is larger. A vent stack serving a toilet must be at least 2 inches in diameter.

Sample Pipe Sizes for Bathroom Fixtures

Fixture	Fixture units	Supply pipe	Drain pipe	Vent pipe
Bathtub	2	½″	1½″	1¼″
Lavatory	1	⅜	1¼	1¼
Shower	2	½	2	1¼
Toilet	4	⅜	1½	2

For supply piping, we recommend either copper tubing or plastic pipe. Copper is accepted by all codes and can be connected to any existing system (see page 34). If you will be installing a shower stall, plan to use flexible copper tubing for the supply lines. Plastic water supply pipe is approved in some areas and not in others (see page 38 for information on plastic piping).

Comparing Drainage and Supply Fittings

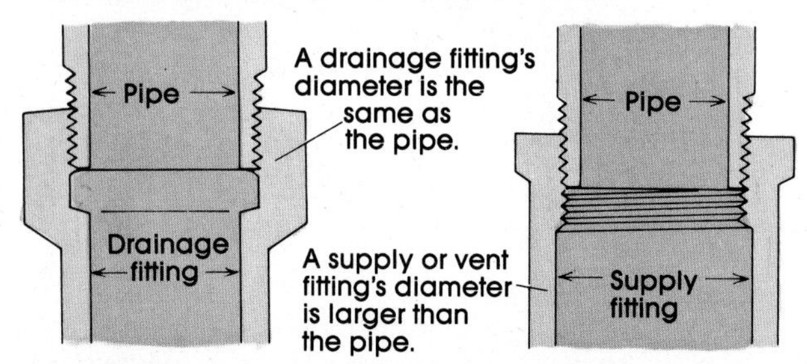

INSTALLING PIPES

Installing pipes in new construction is a matter of doing the various jobs in the right order and making sure the framing accommodates the piping and fixtures. If you have carefully planned your work, you should be able to proceed through the following stages step by step.

Assembling the DWV System

Once the foundation has been poured and the posts and girders are all in place, you can do the initial plumbing. Some plumbers like to wait until the floor joists are in

Placing Joists Beneath a Bathtub

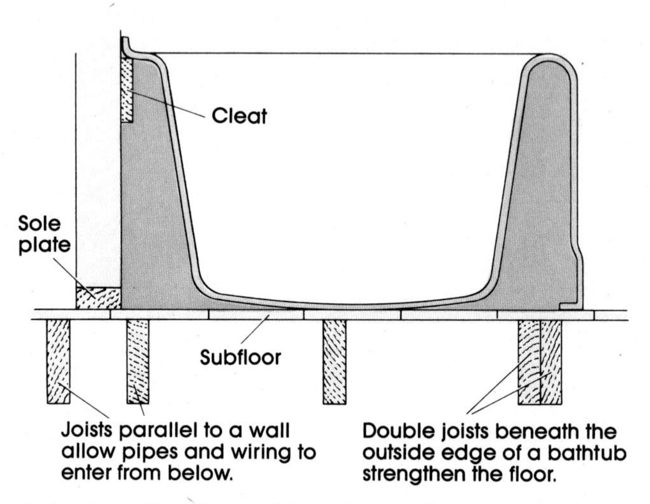

place, too. The floor joists in the bathroom must be set properly to allow room for the pipes to go up into the wall and to avoid the need for cutting through them when installing drain pipes for the tub, shower, and toilet. This means that a joist set parallel to a wall should never be placed directly under the wall. When joists are parallel to the long side of a bathtub, they should be doubled under the outside edge (see drawing).

At this stage, install the first 5 or 6 feet of soil stack with a cleanout. The soil stack must be as close to the toilet as possible. If the building drain and soil stack are of different sizes, you will need an adapter or reducer to connect them (see page 41). Wait to extend the rest of the stack until after the framing of the room is complete.

Now locate the center of the hole for the toilet drain and cut a hole slightly larger than the fitting that will go through it. The closet flange under the toilet (see page 63) will connect to a closet bend that will turn toward the stack. If the fitting does not provide complete support to the soil pipe, you may need to build a support for the stack clamps.

Start assembling the drain pipes at the toilet drain. Suspend the closet bend from the joists or from nailed-on braces. Then work backward to it from the building's drain line in the wall or floor. Horizontal DWV pipe must be supported at specific intervals as designated by your code (see page 46). The drains for the tub and/or shower and the lavatory can also be put in place. Put in any branch tees for drains and revents as needed. Stub out all drains as they can be fitted with fixture traps after the walls have been installed (see page 67).

Installing a Soil Stack Base

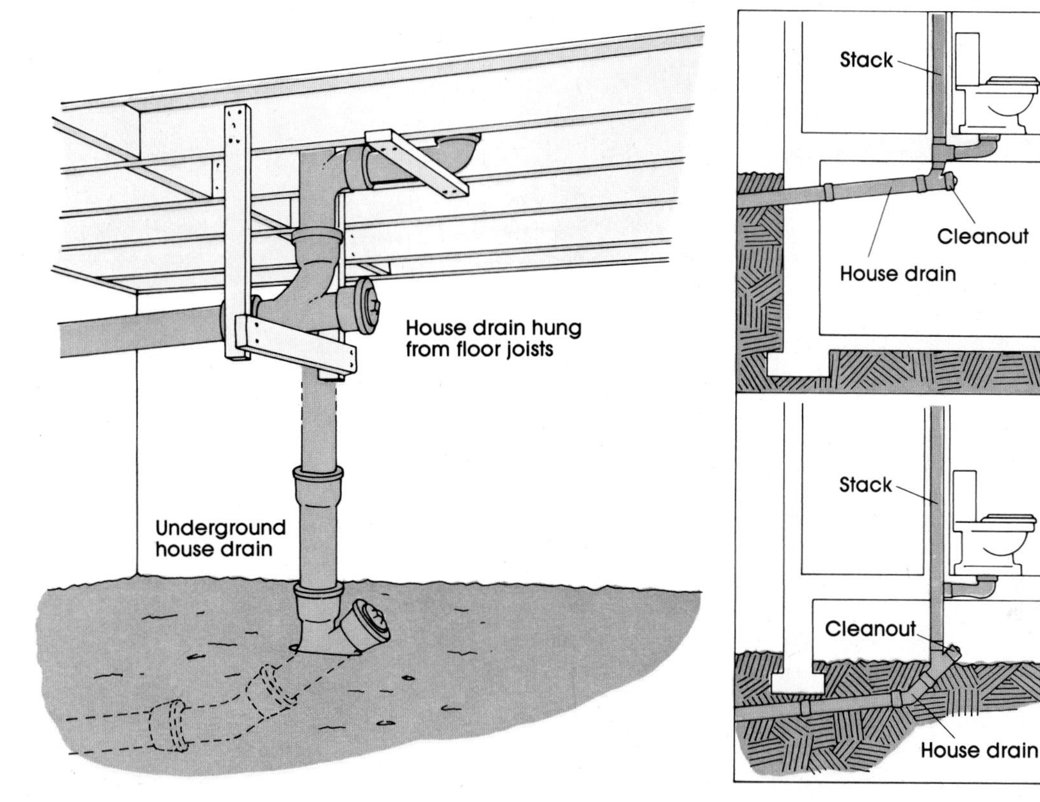

If you are installing a bathtub or lavatory with an overflow, check your local code for regulations. The waste pipe must be designed so water never rises in the overflow and must be connected on the fixture side of the trap (see page 69).

The main drain for a new toilet will have to be vented out the roof. Drains from the tub and lavatory can probably be wet vented.

Running Supply Lines

Run the hot- and cold-water supply lines from their source and extend them upward to a few inches higher than they will eventually be stubbed off. Follow the rough-in dimensions specified for each fixture (see the drawings). Make sure to keep the pipes at least 6 inches apart. Install the main line runs first and then the branch runs, supporting horizontal pipe at intervals of 4 to 10 feet to minimize movement from hammering and temperature changes (see page 46). At the point where branch runs to fixtures come through the wall, install tees. Above each tee, install a 12-inch length of capped pipe for an air cushion chamber, if required.

Once you have completed the rough plumbing (the inside-the-wall work), you must have your work inspected by a local authority. Be sure to arrange to have this done before you close the walls or you will have to open them up again. You may also find that a pressure test is required.

Rough Plumbing Dimensions

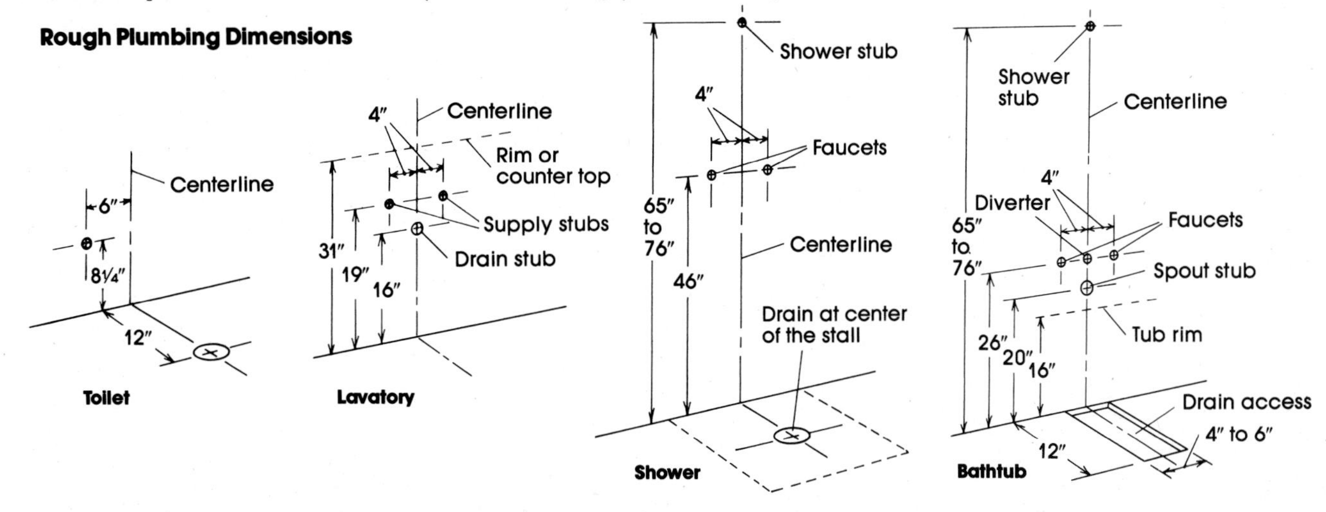

Framing Pipes and Fixtures

When the joists and subfloor are in place, build the wall framing around the pipes. Build the wall enclosing the soil stack with a 2 × 6 or 2 × 8 sole plate, nailing 2 × 4 studs in place flat against the edges of the plate.

In addition to accommodating the framing of your new bathroom to the pipes, you must provide for the attachment of the fixtures:

■ Install a 1 × 8 board across the studs behind the lavatory if it is to be wall-hung, and notch the 1 × 8 into the studs so it is flush. For most fixtures, the distance from the unfinished floor to the top of this board will be 32 ¾ inches. Of course, you can vary this distance according to the height of the people using the fixture.

■ Install a 1 × 4 or 1 × 6 board across the studs to support the shower head, notching the crosspiece in so it is flush. The shower head will be attached to its midpoint—usually about 73 inches above the floor.

■ Framing for a bathtub should include support blocks for the side and ends of the tub that go against the wall and blocking at the rim for drywall nailing. If you need to make room for the pipes at the head of the tub, measure up the wall the height of the tub and saw out the stud closest to the center of the tub. Insert a 2 × 4 between the joists, nail it in place, and cut two vertical studs to support it. Or you can nail a 2 × 4 to the studs to support the tub flange, as shown in the drawing.

■ Any handrails or grips around the tub or toilet also need 2-inch blocking between the studs.

■ For the medicine cabinet, you will need headers at the top and bottom and vertical side nailers. A typical cabinet fits into a 16 × 22-inch opening between the studs, but you can cut out a section of stud to place the cabinet where you want it.

Thickened Wall to Accommodate the Soil Stack

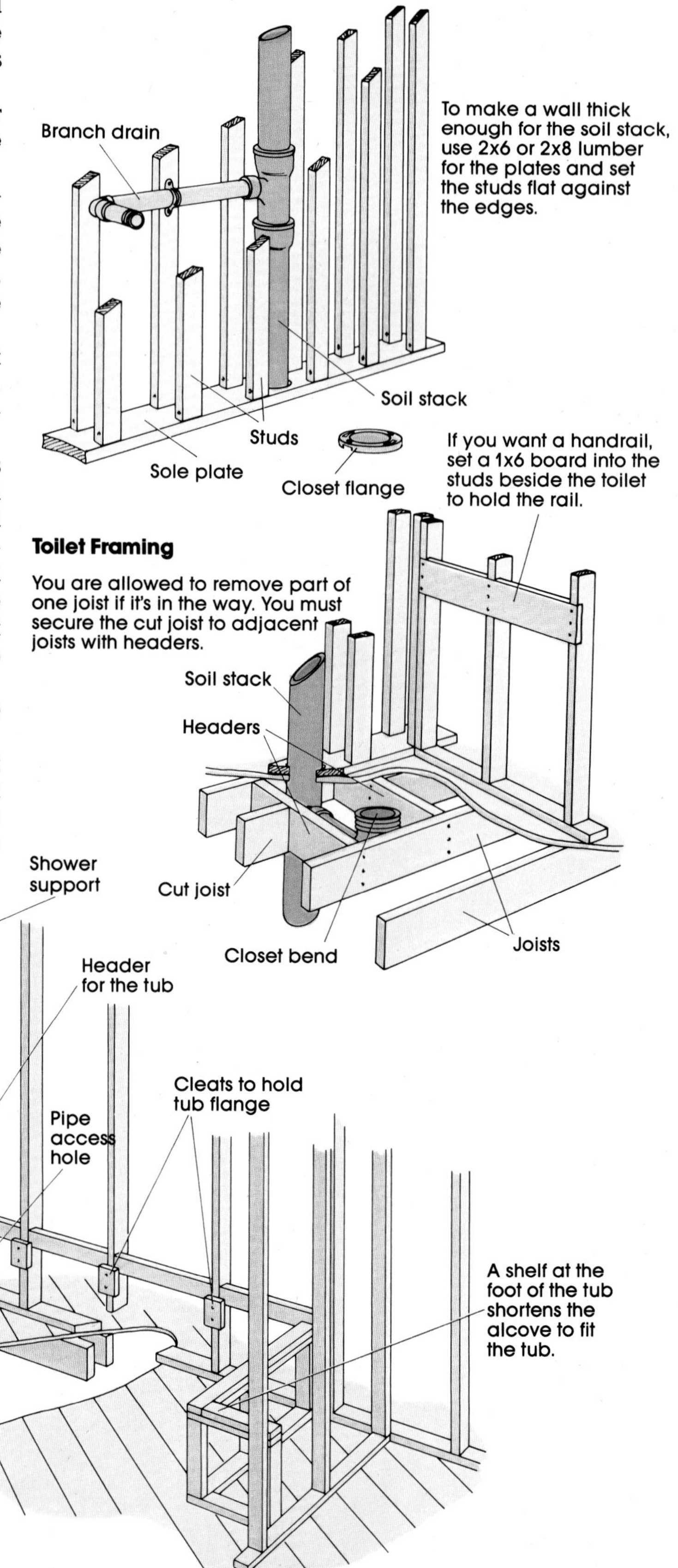

Toilet Framing

You are allowed to remove part of one joist if it's in the way. You must secure the cut joist to adjacent joists with headers.

Framing for Bathroom Fixtures

Headers and framing for medicine cabinet

Support for wall-hung lavatory

Double joist under the edge of the tub

Finishing the Job

Once the framing for the pipes and fixtures is complete, add elbows and stubs to the drain and supply pipes so they will extend through the finished wall in the proper position (see drawings). Install the bathtub (see page 69) and/or shower (see page 71) at this time, too. Check your local code to see if it requires you to provide access to the supply and drain pipes serving your tub. You can usually install an access door in a closet or crawl space.

When the framing of the new room is finished, including the roof, extend the remainder of the soil stack and any other vents to their proper termination 12 inches above the roof line.

Now you can install the drywall or paneling and any built-in cabinets. The final step in the plumbing is the installation of supply stops for each fixture (see page 54), faucets (see page 52), shower stall (see page 71), lavatory (see page 65), and toilet (see page 63).

Exposing Pipes

1. After making test holes and finding the pipes, use a tape measure to find and mark the position of the closest studs.

Marks at the edges of studs

Installing fixtures in a room that already exists can present more problems than starting from scratch. If you want to conceal all the piping inside the existing walls and floors, you will have to remove a lot of drywall or lath-and-plaster wall covering and pry up a lot of flooring. You will also have to cut, drill, reinforce, and maybe replace some floor joists and wall studs to accommodate the soil stack and drainage pipes. Because supply piping is quite small, in most cases it can be run alongside the drain pipes without additional damage to the existing structure.

Exposing the Pipes

Once you have found the existing soil stack, you can use it as a starting point to locate other pipes and fixtures hidden in the walls and floors. You can sometimes track the pipes by turning on the water, one faucet at a time, and listening for the flow with your ear against the wall. You may need to drill test holes to determine the exact location of the hot- and cold-water supply pipes. When you have located the supply lines, turn off the water at the main shutoff valve and drain them through the lowest faucets before you cut an opening in the wall.

As close as possible to where you plan to install your fixture, carefully knock a hole in the plaster or plasterboard to expose the existing plumbing. Insert a steel tape measure into the hole on either side of the pipes until it contacts the closest studs. Mark the location of each stud edge on the wall and outline a large rectangle—about

2. Outline the area you'll need to work, drill starter holes at each corner, and use a knife or keyhole saw to cut the opening.

3 feet high—to form a work area. Drill starter holes in each corner and use a keyhole saw or heavy knife to cut the opening.

Making Connections

Just as in new construction, you will need to connect each fixture to incoming hot- and cold-water lines and to the DWV system. Wherever possible, tap into existing pipes.

Tapping the drain stack. First you must make sure that the stack is solidly anchored. If it slips even slightly, the roof seal at the top may break and cause a leak. To anchor the stack, install additional stack clamps above and below the section to be removed. (A stack or floor clamp consists of two perforated steel straps bolted together at the ends.) Put one strap to the back of the stack and the other in front. Tighten the bolts and support the ends with wood cleats nailed to the sides of the studs.

Measure the height of the new tee or tee Y fitting and mark it on the stack. Use a pipe cutter to cut along the marked lines (see page 29). Slip the rubber sleeves of hubless pipe connectors over the cut ends, positioning them just above and below the pipe opening. Insert the

Tapping the Drain Stack

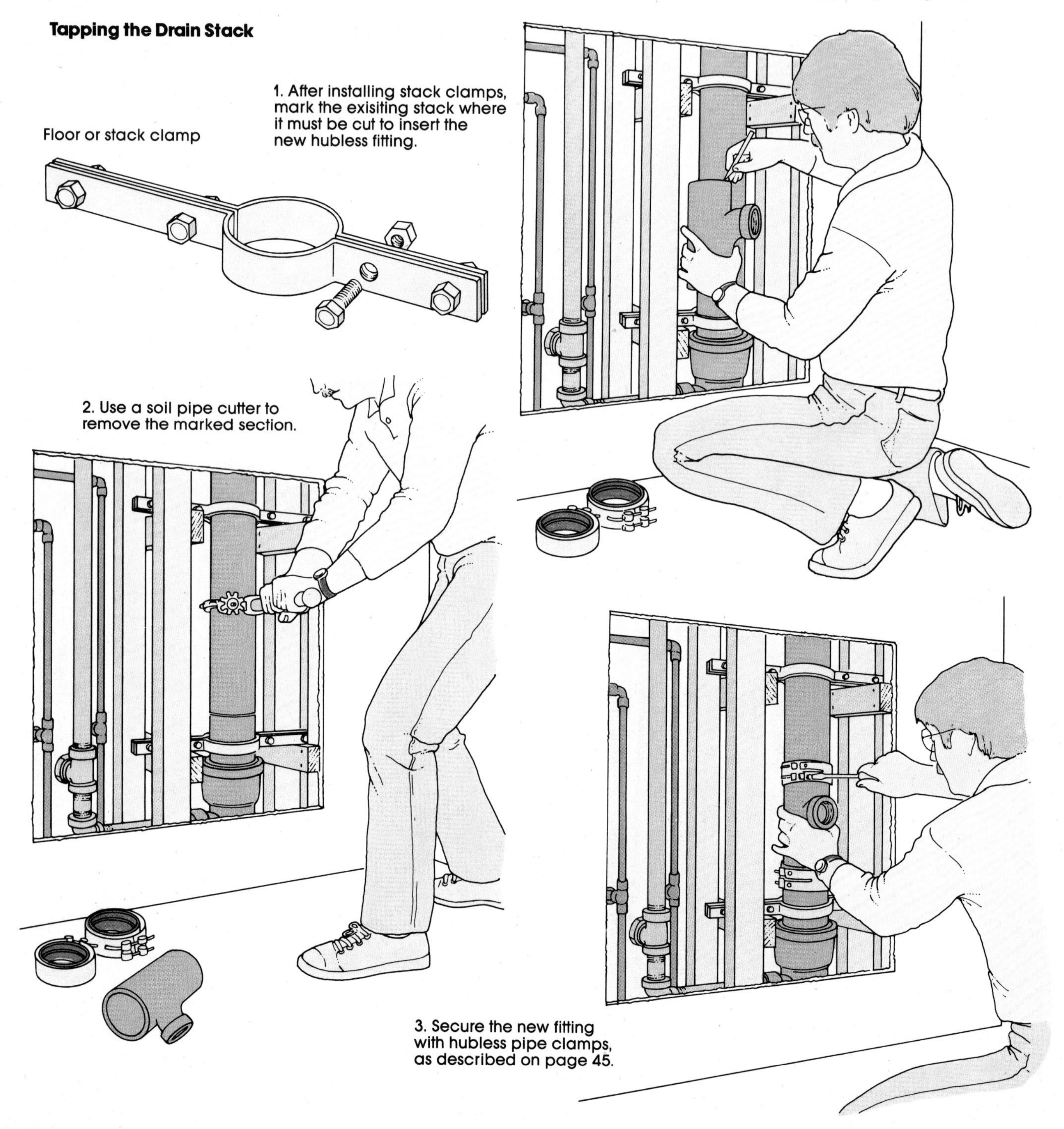

1. After installing stack clamps, mark the exisiting stack where it must be cut to insert the new hubless fitting.

2. Use a soil pipe cutter to remove the marked section.

3. Secure the new fitting with hubless pipe clamps, as described on page 45.

fitting and slide each sleeve over a joint. Place the band clamps over each sleeve and tighten the screws. Leave the support clamps in place.

Tapping supply pipes. Procedures for connecting fixtures to supply pipes vary according to the kind of piping that exists. In all cases, you'll want to install a tee, and if the pipe you want to use differs from the existing pipe, an adapter as well.

If the existing pipe is threaded, iron, or brass, you'll have to use a wrench on the fitting to which it connects. This may involve cutting more than one wall opening. You'll have to cut the pipe; unscrew it from the fittings on each end; and install three pieces of pipe, the new tee, and a union.

If the existing pipe is copper or plastic, you can just cut out a section of pipe and sweat or cement the tee in place. The tee can be installed with or without slip fittings, depending upon the rigidity of the existing pipe (see drawings).

Once the tee is in place, install an adapter if it's needed and extend the piping to the location of the new fixture you want to install.

Tapping Supply Lines

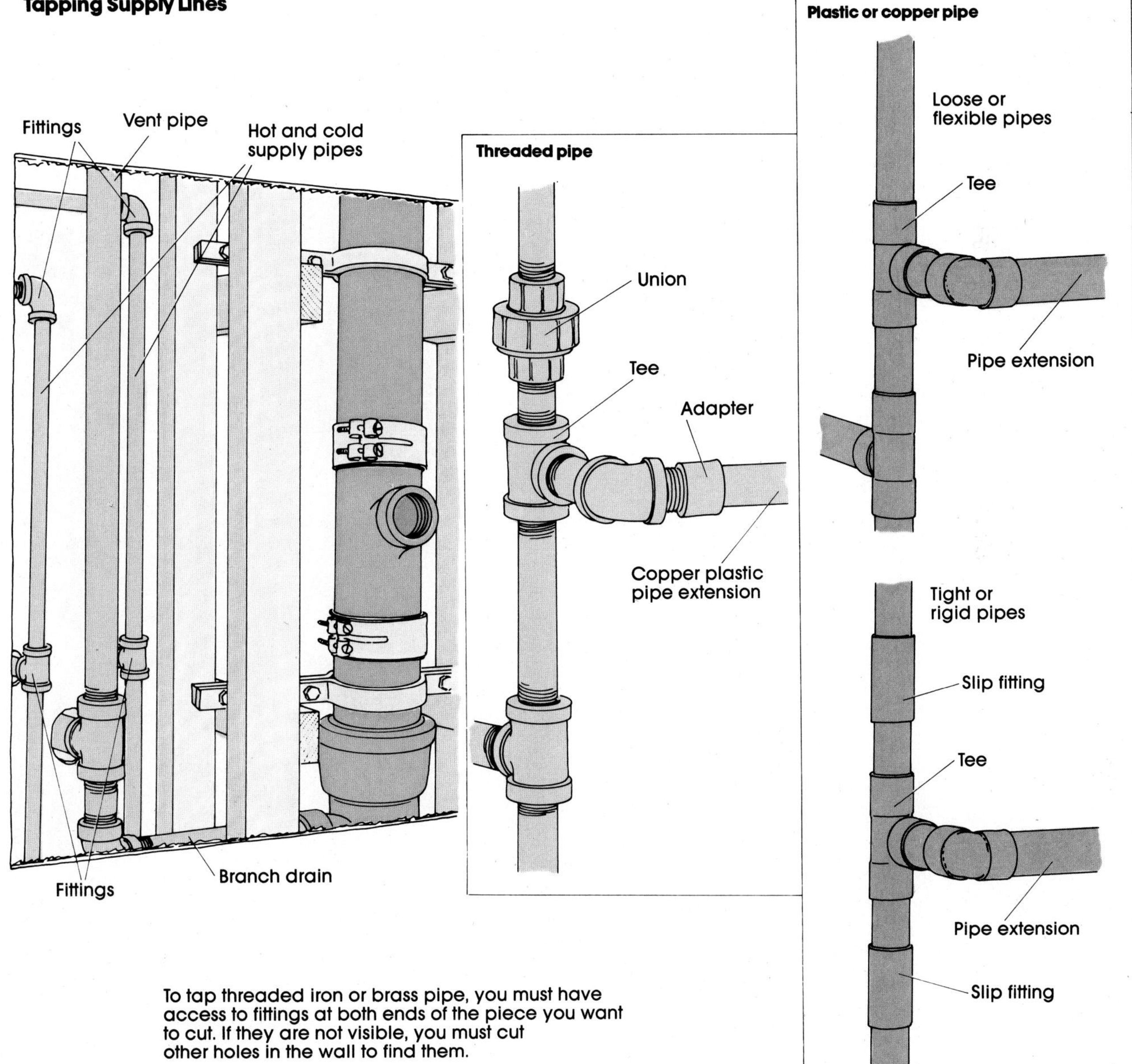

To tap threaded iron or brass pipe, you must have access to fittings at both ends of the piece you want to cut. If they are not visible, you must cut other holes in the wall to find them.

Closing the Opening

In all new connections, leave the wall open for a couple of days to check for leaks before you close the opening. After you have checked your connections, nail cleats against the studs on either side of the opening. Cut a piece of plasterboard to fit the opening and cut slots in the board to fit around the pipes. Nail the plasterboard to the cleats and then close the slots with strips of plasterboard, taped smooth against the wall. Or you can install the plasterboard before you make your connections to the fixture, which will allow you to cut round openings that will fit the pipes.

In most cases, you will need to run a short length of pipe from the fitting through the wall surface. Mark the replacement plasterboard and drill holes for the tubing or pipe, using a wood bit or hole saw. You can patch around the pipe with plaster or taping compound. Install a chrome-plated escutcheon to hide the hole edges.

Closing the Opening

1. Nail cleats to the studs on each side of the opening.

2. Cut plasterboard to fit the opening and drill holes or cut slots for the pipes to come through. Nail the panel to the cleats and seal the edge with drywall tape and patching plaster.

Back-to-Back Sinks

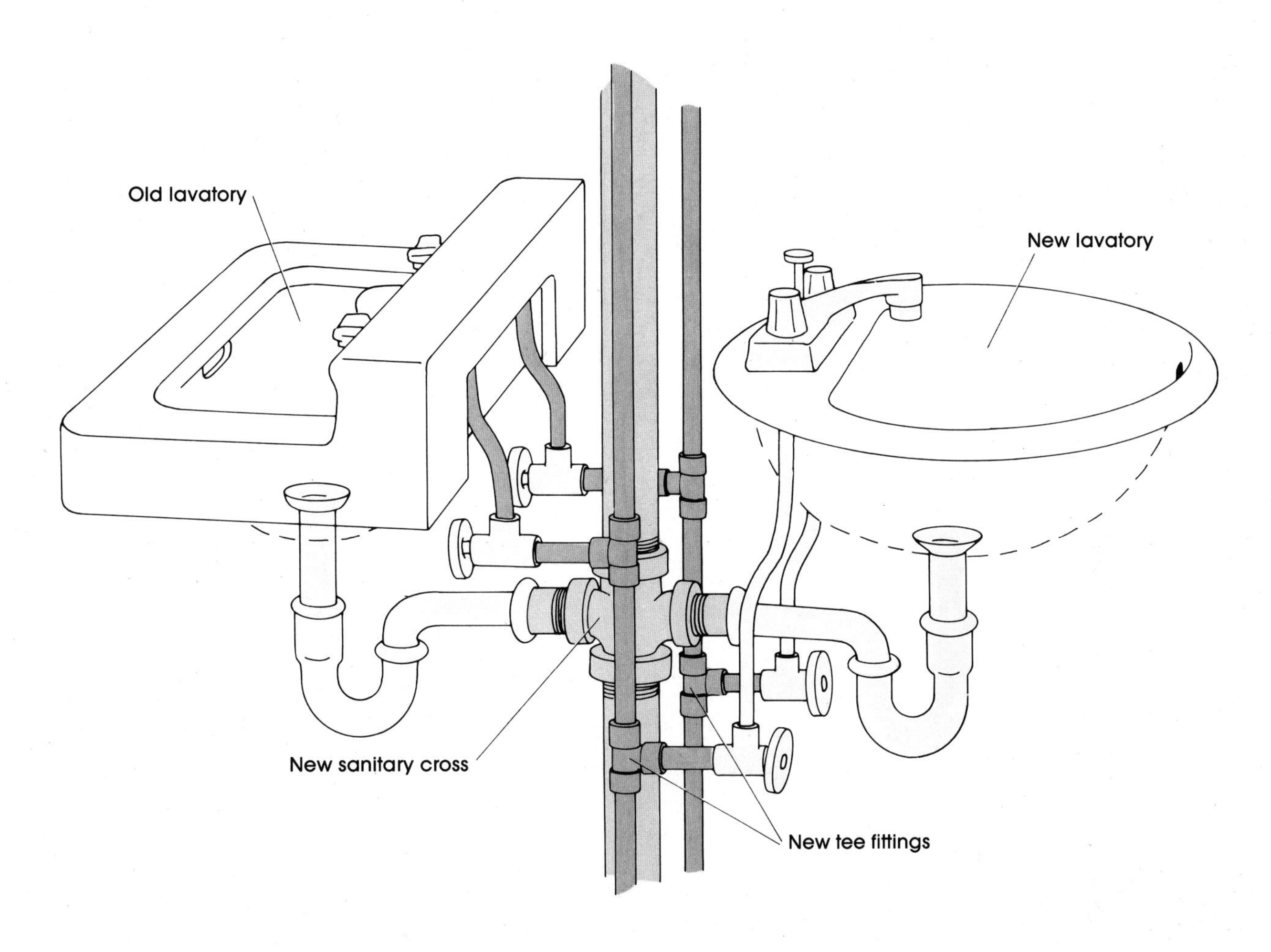

Supporting Fixtures

Each fixture will need to be supported, just as in a new installation (see page 86). This means that to install new fixtures in an existing structure, you will need to open up the wall or floor. Because of its size, install the bathtub first (see page 69), remembering to leave access to the plumbing at the head of the tub if your code requires it.

When installing a toilet in a room not originally designed for it, you may find that you cannot place the closet bend between the joists. If you must remove a section of joist, cut off the end—never a piece from the middle. Then securely fasten the cut joist to the joists on either side with a header, as shown. To be safe, shore up the floor before starting this process.

Adding a Lavatory

You may be able to extend piping from an existing fixture to serve a second fixture. There are several ways to do this, but one precaution pertains to all of them. Make sure the new drain pipe enters the existing stack at a point low enough to make the waste flow downhill, but not so low it will create suction on the new fixture's trap. (See page 84 for the maximum distances drain pipes can travel from the fixture trap to the stack.) Also, check your local code to be sure a separate vent pipe is not required.

Side-by-side sinks. To connect a second lavatory to one already in the room, you can make all the connections outside the wall. Attach the new fixture so that its drain hole is no more than 30 inches away from that of the existing sink and no more than 6 inches higher—although the new drain must start somewhat higher than the old one to provide some slope. Cut the existing supply lines and install tee fittings to supply water to the new sink. Do the same for the drain line. Remove the tailpiece of the existing fixture and install a slip-joint tee above its trap. Connect this tee to the drain holes of both fixtures with a tailpiece for the existing sink and a tailpiece and 90-degree slip elbow for the new sink. Remove the existing shutoff valves, install tees behind them, and replace the existing shutoffs. Then run piping from the tees to the shutoffs of the new sink (see page 85).

Sinks connected through a wall. If you want to connect a new lavatory to an existing one in an adjacent room,

Side-by-Side Sinks

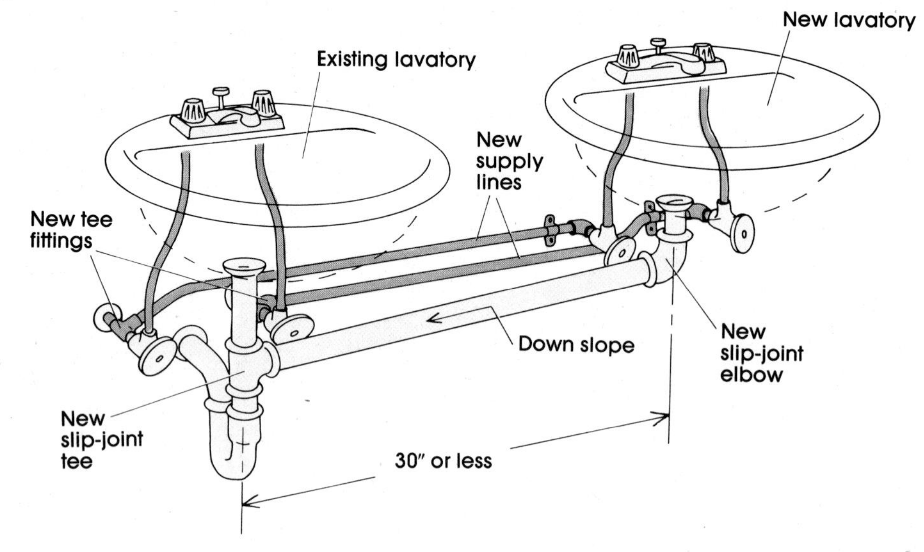

you will need to open the wall to extend the supply and drain lines. If the sinks will be back to back, the drain and supply lines should be kept as short as possible. Replace the old stack fitting with a sanitary cross (which has no internal ledges that trap waste). Make the supply connections by installing new tee fittings on the existing supply lines and run pipes to the shutoffs on the new fixture.

If the new sink will not be directly behind the old one, you will need to extend the supply and drain lines along the wall. To run a drain pipe along a wall, use a 90-degree elbow and a spacer. Temporarily support the pipe with pieces of lumber or bricks until you connect the pipe. If the pipe extends along the wall for more than a few feet (see page 46), use metal straps to anchor it to the wall.

Concealing Pipes in Existing Walls

To hide plumbing inside a wall or between a ceiling and the floor above, there must be enough clear space for the pipe and the fittings. Two-inch cast iron pipe and fittings (4 inches wide) will not fit inside a standard wall made of 2×4 studs (which are 3½ inches wide). This same wall, however, has plenty of space to contain 2-inch plastic or 3-inch copper pipe and fittings.

Because plastic, copper, and threaded iron pipe can be cut to virtually any length and the fittings placed wherever you want them, it is possible to plan your job so a length of pipe without fittings goes through an especially narrow area. You can put on fittings where more space is available.

If an existing wall has too little space for the piping that must go into it, enlarge the wall by setting new studs against the old ones and making the wall thicker. Even a 6-inch or larger addition to the thickness of an interior wall is difficult or impossible to detect once the wall surface is replaced and painted.

When running pipes horizontally inside a stud wall or across joists in a floor, you must drill or notch the studs or joists. In order not to weaken the wall, you may have to set new studs next to the old and thicken the wall even though there is room enough for the pipes.

Here are some regulations based on national building codes to prevent weakening the structure of your house too much:

- Do not notch a stud deeper than one third of its thickness unless you reinforce it with a steel strap.
- Do not notch deeper than one half of the stud, even if you do reinforce it.
- Never notch a joist in the center half of its length. Make any notches in the quarter of the length at each end.
- Never notch deeper than one quarter the height of a joist and always reinforce the notch with a steel strap or 2×2 board.
- Instead of a notch, a hole can be drilled anywhere along the length of a joist as long as it is centered from top to bottom and its diameter does not exceed one quarter the height of the joist.

Especially with small-diameter supply pipes, you can often minimize or entirely eliminate notching by lifting the flooring and cutting a groove in the subfloor. You may need to add furring strips between the subfloor and the flooring, as shown in the drawing.

Concealing Pipes in the Floor

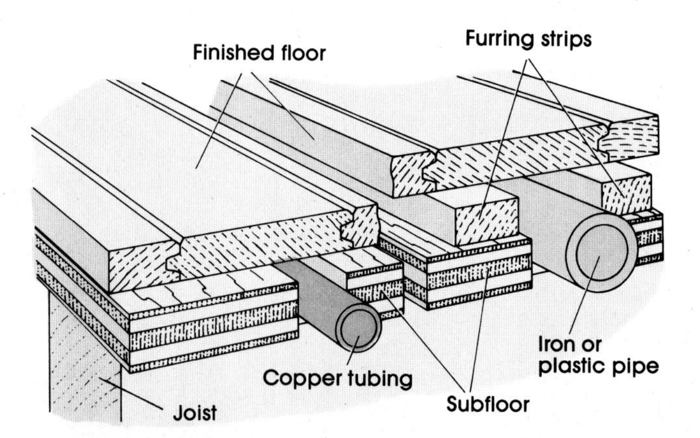

GLOSSARY

ABS (acrylonitrile-butadine-styrene). Rigid black plastic pipe, used for drain, waste, and vent systems.

Adapter fitting. Device connecting pipes of different materials (plastic or metal) or sizes.

Aerator. Snap-on or screw-on device for a faucet; fills flowing water with air bubbles and controls splashing.

Air chamber. Vertical length of capped pipe extending from the supply pipe, where it enters the fixture; supplies an air cushion to prevent water hammer.

Air gap. Vertical distance between the lowest opening of a supply pipe or faucet and the highest flood rim of a sink; prevents back-siphonage of wastewater into the supply lines. Also, the open space through which a pump-drained appliance (dishwasher, washing machine, and so forth), must discharge its wastewater before that water enters the drainage system; prevents excessive pressure and siphoning within the system.

Allen wrench. Hexagonal, rod-shaped tool for turning Allen screws.

Auger, closet. Coiled wire device with a crank, housed in a bent tube to protect the toilet surface; used to unclog toilet drains.

Auger, drain. Coiled wire device with a crank and hooked end for unclogging sink drains.

Ball-cock. Automatic toilet-tank water supply valve, controlled by a float ball, for bringing water into the tank; usually of the antisiphon type.

Basin wrench. Tool for removing or installing nuts or bolts in hard-to-reach places, usually under sinks.

Branch. Any part of a pipe system connected to a fixture.

Building drain. The lowest horizontal drain pipe in the building; carries all waste to the sewer.

Bushing. Device used to adapt a pipe, fitting, faucet, or valve to a threaded opening of a larger size. Has internal and external threads.

Cap. Cover used to close off the end of a pipe.

Cartridge. The replaceable unit in some washerless faucets; controls the flow of water.

Cast iron pipe. Large, heavy pipe (hub-and-spigot or hubless) used for drain, waste, and vent lines, and sewer laterals.

Caulking. The sealing of bell-and-spigot drain pipe joints with oakum and lead.

Chain wrench. Adjustable wrench in which a chain is used for gripping. Best for use with large pipes or where other wrenches won't reach.

Channel-lock pliers. Large pliers with adjustable, parallel jaws.

Check valve. Valve that lets water flow in one direction only.

Clamp straps. Straps used to hold or support pipes.

Cleanout plug. Easy-to-reach opening in the drainage system; used in removing obstructions.

Close nipple. Very short length of externally threaded pipe on which the threads from each end meet or almost meet in the middle.

Closet (water closet). Toilet.

Closet bend. Pipe connecting the waste exit of the toilet bowl to the soil-stack fitting.

Closet spud wrench. Tool used to loosen or tighten closet spud nuts, strainers, and trap nuts.

Copper pipe. Thickwalled, rigid pipe made of copper, found in older homes.

Copper tubing. Soft, pipelike material sold in pipe lengths and rolls.

Coupling. Fitting used to connect two or more pipes, or a pipe to a fixture.

CPVC (chlorinated polyvinyl chloride). Chemical material used to make plastic pipe.

Cross connection. Physical link between contaminated water or sewage and fresh water.

Die, pipe. Tool for cutting threads on the outside of pipe.

Die stock. Tool that holds the pipe die.

Diverter. Valve for changing the flow of water from one faucet to another.

DWV (drainage, waste, and vent). Refers to all or part of the drain-waste-vent system.

Effluent. Liquid discharged from septic tank.

Eighth bend. Bent fitting that connects pipes to each other at a 45-degree angle.

Elbow (bend, ell). Fitting that connects pipes at an angle to each other.

Ell. *See* Elbow.

Escutcheon. Ornamental plate for covering a wall opening where pipe penetrates.

Faucet. Valve for controlling water flow; generally used with a fixture such as a sink.

Female threads. Any pipe threads located internally.

Finish plumbing. Attractive visible plumbing fixtures or pipe.

Fitting. Device that connects pipe to pipe or fixture.

Fixture. Any device permanently attached to the running water system of a house (for example, sink or toilet).

Fixture branch. Drain that discharges into a drain or stack and serves one or more fixtures.

Flange. Extending rim or edge at one end of a pipe shaft that gives support or a finished appearance.

Flange, closet (floor flange). Fitting that connects a toilet to the floor and the closet bend.

Floor and ceiling plates. Decorative fitting for the floor and ceiling where pipes enter them.

Flow pressure. Pressure within the water supply pipe by the faucet or water outlet while water is flowing.

Force cup (plumber's friend, plunger). Rubber cup, usually with a wooden handle, used to dislodge clogs in sink drains and toilets.

Galvanic action. Electrical process whereby metallic elements are leached from one part of a system and attracted to another.

Gasket. Flat device, usually made of fiber or rubber, used to provide a watertight seal between metal parts of joints.

Ground. Safety wire, often attached to plumbing from the electrical system in the home. Usually carries no current unless a short circuit occurs.

Hot-water main. Primary pipe that conducts hot water to various fixtures in the home.

Hubless pipe. Pipe with smooth ends (no spigot or hub).

Inlet opening. Aperture through which a liquid or gas enters a fitting.

Joint runner. Collarlike device that keeps molten lead in place while sealing a joint in a horizontal run of cast iron pipe.

Lead. Heavy metal used to seal cast iron pipe.

Main drain. Line that collects discharge from branch waste lines and carries it to outer foundation wall, where it connects to the sewer line.
Main vent (stack). Main artery in the venting system to which vent branches may connect.
Male threads. External threads cut in the ends of pipes and fittings.
Nipple. Short length of pipe, externally threaded at both ends.
Non-potable water. Contaminated water unsafe for drinking, washing, or cooking.
Packing. Fibrous or rubberlike material that prevents leakage around valve or faucet stems.
Packing nut. Nut that holds the stem of faucet in position and contains or confines the packing material.
Phillips head. Type of screw and screwdriver with an X-shaped, rather than conventional slotted head.
Pipe joint compound. Material used as lubrication or sealer on threads of pipes and fittings.
Pipe vise. Special vise made to hold pipes.
Pitch. Downward slope of a drain pipe in the direction of liquid flow.
Plumber's friend. *See* Force cup.
Reamer (tapered reamer). Tool for smoothing and removing burrs from the ends of freshly cut pipe.
Reducer. Fitting enabling pipes of different diameters to fit together.
Reducing tee. Fitting for connecting three pipes, either one or all of which are of different sizes.
Relief valve. Valve used on water heaters to prevent excessive heat or pressure in the tank.
Rough-in. Installation of the part of the plumbing system (water supply, drainage, and vent piping and necessary fixture supports) that can be finished before fixtures are installed.
Saddle tee. Fitting that taps into the side of a pipe, used to make a quick connection to an existing water line.
Sanitary fitting. Drain fitting with a smooth interior and connections that allow solid material to pass through without clogging.
Septic tank. Watertight receptacle that receives the raw sewage from a house sewer.
Service entrance line. Pipe connecting the water company piping to the water meter.
Service tee. Tee fitting with male threads on one run opening and female threads on the other two openings.
Shutoff valve. Fitting for shutting off water flow to a pipe or fixture.
Side vent. Vent connecting to a drain pipe at an angle of 45 degrees or less.
Single-lever faucet. Faucet with one handle that controls the flow and mixing of hot and cold water.
Snake (auger). Long flexible cable, used to dislodge clogs in pipe.
Soil stack. *See* Main drain.
Solar collector. Panellike or tubular device that absorbs the sun's heat and transfers it to water and/or air, thus, heating them.
Stack. Main vertical pipe in the DWV system; usually extends from the basement to a point above the roof.
Stem. Part of the faucet; holds the handle at one end and the washer at the other.
Stop valve. Fitting, usually located under a fixture, that stops the flow of water to that fixture.
Straight cross. Fitting that connects four pipes of the same diameter.
Straight tee. Fitting that connects three pipes of the same diameter.
Strap wrench. Tool with metal body and heavy cloth or metal mesh belt; used for turning pipes.
Street tee. *See* Service tee.
Sweating. What happens when moisture condenses on the outside of pipes or toilet tanks. Also, another term for sweat soldering.
Sweat soldering. Method of connecting copper tubing with solder and a propane torch.
Tee. Pipe fitting in the shape of a T, with three openings for the connection of pipes.
Trap. U-shaped device that allows water and sewage to flow through it while blocking the flow of air and gas from the other direction.
Union. Three-piece fitting for joining pipes.
Valve. Device that controls the flow of liquid or gas through or from a pipe.
Vent. Pipe that provides the flow of air into and gases out of a DWV system and prevents siphoning of water from traps.
Vent stack. Vent pipe that is set up vertically.
Water hammer. Pounding or knocking sound in water pipes, due to a sudden change in the pressure when a faucet is shut off.
Wax ring. Donut-shaped seal, made of wax, used at base of toilet to prevent leakage.
Wet vent. Drain or waste pipe that also acts as a vent for one or more fixtures on the same line.
Wye branch. Fitting connecting three pipes, usually in the DWV system, allowing one pipe to enter it at less than a 90-degree angle.
Y-branch. *See* Wye branch.

INDEX

NOTE: *Italicized* page numbers refer to illustrations.